IS IT POSSIBLE IN UNIVERSE?

ANSWERS TO EVERYTHING IN UNIVERSE

DEBARPAN PAUL

This book is dedicated to every curious mind that wants to explore the boundless possibilities of the universe. May it serve as a beacon of knowledge, enlightening and inspiring individuals from all walks of life. Let it be a reminder that the quest for understanding is a journey we all share.

Contents

Contents

Preface

In an era where the boundaries of human knowledge are continually expanding, we find ourselves ever more fascinated by the mysteries of the universe. This book, "Is It Possible in the Universe?", is a humble contribution to the quest for understanding the vast, enigmatic cosmos that surrounds us.

From the smallest subatomic particles to the largest galactic structures, the universe is a tapestry woven with intricate patterns and phenomena that challenge our comprehension. This book seeks to bridge the gap between the known and the unknown, presenting complex scientific concepts in a manner that is accessible to all.

Throughout these pages, you will embark on a journey through space and time, exploring the possibilities that the universe holds. We will delve into questions that have puzzled humanity for centuries: Are we alone in the cosmos? What is the nature of dark matter and dark energy? Is our universe just one of many in a multiverse?

This book is for everyone—the curious student, the avid reader of science fiction, the lifelong learner, and anyone who has ever looked up at the night sky with wonder. It aims to ignite your imagination and provide you with knowledge that is both profound and practical.

Each chapter is a step forward in our collective journey to understand the universe. I hope that as you read this book, you will find not just answers, but also new questions to ponder. For it is in questioning that we continue to learn and grow.

Welcome to a voyage of discovery. Welcome to the infinite possibilities of the universe.

Foreword

In the boundless tapestry of the universe, where each star is a luminous thread and each galaxy a grand motif, lies a world of infinite possibilities. This book, "Is It Possible in the Universe?", embarks on a profound journey to explore these possibilities, weaving together the known and the unknown, the scientific and the speculative, into a cohesive narrative that invites readers to question, to ponder, and to dream.

As you delve into the pages that follow, you will encounter a diverse array of topics that stretch the imagination and challenge the boundaries of our understanding. From the enigmatic nature of dark matter and dark energy to the tantalizing prospect of extraterrestrial life, from the philosophical implications of living in a multiverse to the scientific wonders of distant exoplanets, each chapter is a doorway into a different facet of the cosmos.

This book is not just for the seasoned astronomer or the avid science enthusiast; it is for everyone. It seeks to democratize knowledge, making the wonders of the universe accessible to all who dare to look up at the night sky and wonder what lies beyond. It is for the dreamers, the seekers, and the perpetual students of life who believe that the pursuit of knowledge is a journey without end.

In writing this foreword, I am reminded of the words of Carl Sagan, who said, "Somewhere, something incredible is waiting to be known." This book stands as a testament to that belief, encouraging readers to embark on their own voyages of discovery and to embrace the infinite possibilities that the universe holds.

As you turn each page, may you find inspiration and wonder, and may your curiosity lead you to new horizons. Welcome to a journey through the cosmos, a journey that is as much about the universe within us as it is about the universe around us.

I have tried to present my thougts by this book
Enjoy the exploration.
[Debarpan Paul]

Acknowledgements

I would like to extend my heartfelt gratitude to all those who have supported and guided me throughout the journey of writing this book.

First and foremost, I would like to express my deepest thanks to Notion Press for publishing my book. Your professionalism and dedication have been instrumental in bringing this work to life, and I am truly grateful for your support.

A special thanks goes to my parents, Debdutta Paul (my father) and Pinki Paul (my mother), for their unwavering support, encouragement, and belief in me. Your love and guidance have been my greatest source of strength.

I am also deeply thankful to all of my teachers and my school, D.A.V. Fulbari Siliguri. Your teachings and support have played a crucial role in shaping my knowledge and perspective, enabling me to embark on this literary journey.

Thank you all for your invaluable contributions. This book would not have been possible without your support.

[Debarpan Paul]

Introduction

Introduction

In the vast expanse of the cosmos, where the glow of distant galaxies intertwines with the dark matter of the unknown, lies the essence of human curiosity. This book embarks on a journey through space, science, and the unfathomable mysteries of the universe, seeking to unravel the enigmatic tapestry that has both fascinated and confounded humankind for millennia.

Our story begins with the birth of the universe, a cataclysmic event that set into motion a cosmic ballet of stars and planets, black holes and quasars. From the fiery crucible of the Big Bang to the serene beauty of a starlit sky, each chapter delves into the intricate workings of the cosmos, blending scientific rigor with the awe-inspiring wonders of the celestial realm.

As we traverse through the chapters, we will explore the frontiers of space, where astronauts and astronomers alike strive to uncover the secrets of distant worlds and alien phenomena. We will delve into the theoretical underpinnings of modern astrophysics, examining the fundamental forces that govern our universe and the groundbreaking discoveries that challenge our understanding of reality.

But this journey is not merely about the cold, hard facts of science; it is also a quest to comprehend the deeper mysteries that lie beyond the reach of our telescopes and instruments. What is the nature of dark matter and dark energy, the unseen forces that comprise most of the universe? Are we alone in the cosmos, or do other sentient beings gaze at the stars from worlds light-years away? And what is the ultimate fate of our universe, a question that has both scientific and philosophical implications?

Join me as we embark on this voyage through space and time, blending the meticulous observations of science with the boundless imagination of human inquiry. Together, we will explore the grand narrative of the universe, a story written in the light of distant stars and the silent echoes of cosmic events long past. This is a journey not just to the farthest reaches of space, but to the very heart of what it means to seek, to question, and to wonder.

Welcome to the mystery of the universe.

CHAPTER ONE

WHAT IS THE UNIVERSE?

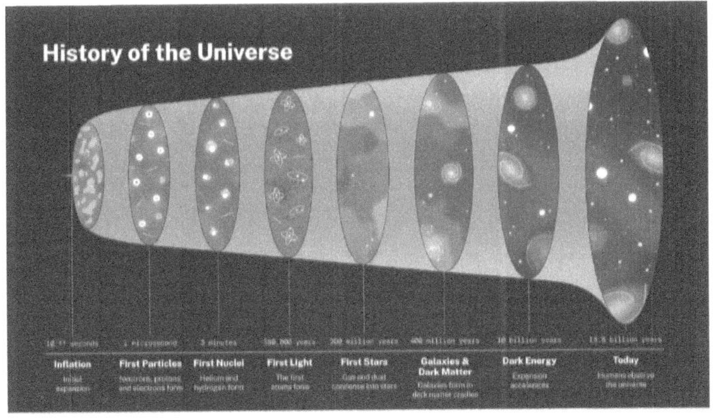

HISTORY OF UNIVERSE

Introduction

The universe is an expansive and intricate tapestry of matter, energy, and space-time that encompasses all known physical existence. From the smallest subatomic particles to the vast galactic superclusters, the universe is an all-encompassing realm where the laws of physics govern the behavior of everything within it. Understanding the universe has been a central pursuit of humanity since ancient times, and it continues to inspire awe and curiosity in modern scientific exploration.

The Scale of the Universe

Cosmic Dimensions

The universe is vast beyond human comprehension. It is estimated to be about 93 billion light-years in diameter, a figure that stretches the limits of imagination. A light-year, the distance light travels in one year, is approximately 5.88 trillion miles (9.46 trillion kilometers). This staggering scale means that the observable universe is an almost unfathomable expanse filled with countless galaxies, stars, and other celestial bodies.

The large-scale structure of the universe resembles a cosmic web, with galaxies distributed along filaments and clusters, separated by vast voids of nearly empty space. At the grandest scales, the universe appears homogeneous and isotropic, meaning it looks the same in all directions and at all locations. This uniformity is a foundational principle in cosmology, known as the Cosmological Principle.

Galaxies and Stars

Galaxies are the fundamental building blocks of the universe. They come in various shapes and sizes, including spiral, elliptical, and irregular forms. Our home galaxy, the Milky Way, is a barred spiral galaxy containing over 100 billion stars. Each star within a galaxy can host planetary systems, like our solar system, making the potential for diverse forms of matter and life almost limitless.

The Birth of the Universe

The Big Bang Theory

The most widely accepted explanation for the origin of the universe is the Big Bang theory. According to this theory, the universe began as a singularity approximately 13.8 billion years ago. This singularity was an infinitely small, hot, and dense point that suddenly expanded, leading to the formation of space and time as we know it.

Early Universe

In the moments following the Big Bang, the universe was a hot, dense plasma of elementary particles. As it expanded, it cooled, allowing these particles to combine and form simple atoms, primarily hydrogen and helium. This period, known as recombination, led to the formation of the Cosmic Microwave Background (CMB) radiation, a relic of the early universe that provides crucial information about its origins and composition.

Evolution of the Universe

Formation of Structures

Over billions of years, the slight density fluctuations in the early universe, influenced by gravity, led to the formation of more complex structures. Gas clouds collapsed to form stars, which then grouped together to create galaxies. These galaxies merged and interacted, forming clusters and superclusters, shaping the large-scale structure observed today.

Stellar Evolution

Stars are the forges where elements heavier than hydrogen and helium are created. Through nuclear fusion, stars produce energy and synthesize new elements, which are then distributed throughout the universe when stars die, especially in spectacular supernova explosions. These processes enrich the interstellar medium with the building blocks for planets and life.

Dark Matter and Dark Energy

One of the most intriguing aspects of the universe is the presence of dark matter and dark energy. Dark matter, which does not emit or absorb light, exerts gravitational forces that influence the motion of galaxies and galaxy clusters. Dark energy, on the other hand, is thought to be responsible for the accelerated expansion of the universe. Together, dark matter and dark energy constitute about 95% of the total mass-energy content of the universe, with ordinary matter making up only about 5%.

The Future of the Universe

Expansion and Fate

The ultimate fate of the universe depends on its rate of expansion and the properties of dark energy. Current observations suggest that the universe will continue to expand indefinitely, leading to scenarios such as the "Big Freeze," where galaxies move so far apart that stars exhaust their fuel and the universe becomes dark and cold. Other possibilities include the "Big Rip," where the expansion accelerates to the point that all structures are torn apart, and the "Big Crunch," where the universe could eventually collapse back into a singularity, although this is less supported by current data.

Multiverse Hypothesis

Some theories propose the existence of a multiverse, where our universe is just one of many universes with different physical laws and constants. This idea, while speculative, arises from certain interpretations of quantum mechanics and string theory, suggesting that our universe might be part of a larger, more complex reality.

Conclusion

The universe is a dynamic and ever-evolving entity, filled with mysteries and wonders that continue to captivate scientists and laypeople alike. From its explosive beginnings in the Big Bang to the potential scenarios of its distant future, the study of the universe encompasses a wide range of disciplines, including astronomy, physics, and cosmology. As we continue to explore and understand the cosmos, we uncover deeper insights into the fundamental nature of reality itself. The journey of discovery is far from over, and each new finding brings us closer to answering the profound questions about our place in the grand scheme of the universe.

CHAPTER TWO

Exploring the Mysteries Before the Big Bang: Speculations and Theoretical Frontiers

The question of what existed before the Big Bang stands as one of the most profound enigmas in cosmology, challenging our understanding of the universe's origins and the nature of existence itself. While our current scientific theories and observational data provide invaluable insights into the evolution of the cosmos, they are limited in their ability to penetrate the veil of cosmic prehistory. Nonetheless, cosmologists and theoretical physicists have put forward a multitude of speculative ideas and hypotheses to grapple with this cosmic conundrum, each offering a unique perspective on the cosmic dawn before the Big Bang.

1. Singularity and the Limitations of Physics

According to the prevailing model of the Big Bang, the universe began as an infinitely dense and hot point known as a singularity. At this moment of cosmic inception, the laws of physics as we understand them break down, and our current theoretical frameworks struggle to provide meaningful descriptions of what may have existed before this epochal event. Some theories suggest that time itself may have emerged with the Big Bang, rendering the concept of "before" an abstraction beyond our comprehension.

big bang

2. Multiverse and the Tapestry of Reality

One speculative idea that has gained traction in recent decades is the concept of the multiverse – a vast ensemble of universes, each with its own unique properties and evolutionary trajectory. Within this framework, the Big Bang may have been just one event among many, occurring within a larger tapestry of cosmic domains. Before the Big Bang, other universes or different regions of spacetime with distinct physical laws and dimensions may have existed, each governed by its own set of rules.

3. Eternal Inflation and Cosmic Birth

Eternal inflation, a hypothesis derived from inflationary cosmology, proposes that the universe undergoes periods of exponential expansion, giving rise to an ever-expanding multiverse. In this scenario, new universes continually bubble forth from the inflating spacetime, each experiencing its own cosmic history and evolutionary path. Before the Big Bang of our observable universe, there may have been other epochs or structures, potentially including previous cycles of inflation and contraction.

4. Quantum Cosmology and the Quantum Foam

The marriage of quantum mechanics and general relativity at the singularity of the Big Bang presents a formidable challenge to cosmologists. Quantum cosmological models speculate about the existence of a pre-Big Bang state governed by quantum principles, where the universe may have existed in a state of primordial flux or quantum indeterminacy. Concepts such as the "quantum foam" suggest a chaotic and fluctuating landscape where spacetime itself undergoes quantum fluctuations on microscopic scales.

5. Cyclic Universe and Cosmic Recurrence

Cyclic cosmology proposes a cyclical model of cosmic evolution, wherein the universe undergoes an endless cycle of expansion and contraction. Each cycle begins with a Big Bang, giving rise to a new epoch of cosmic evolution, followed by a period of contraction culminating in a Big Crunch. Before each Big Bang, there may have been a previous contracting phase of the universe, suggesting a cyclical and recurring nature to cosmic history.

6. Philosophical Implications and Epistemological Boundaries

Beyond the realm of scientific inquiry lie profound philosophical reflections on the nature of time, causality, and the limits of human knowledge. The question of what existed before the Big Bang raises fundamental questions about the nature of existence itself – is the cosmos a finite entity bounded by temporal constraints, or does it partake in an eternal and timeless reality beyond our comprehension? Moreover, the very act of speculating about the pre-Big Bang epoch underscores the inherent limitations of human cognition and the boundary between scientific inquiry and metaphysical speculation.

In conclusion, the mystery of what existed before the Big Bang remains one of the most tantalizing puzzles in cosmology, inspiring ongoing research, speculation, and philosophical contemplation. While our current scientific understanding offers valuable insights into the evolution of the universe, the true nature of cosmic prehistory may forever elude our grasp, inviting us to explore the frontiers of human knowledge and imagination in our quest to unravel the mysteries of the cosmos.

CHAPTER THREE

WHAT IS TIME?

"What is time?" This question, as simple as it seems, has perplexed philosophers, scientists, and thinkers for centuries. Time, the invisible force that governs our lives, seems to both define and elude our understanding. In our quest to comprehend its essence, we embark on a journey into the depths of this enigmatic concept.

The Illusion of Linearity

Our perception of time is deeply ingrained with the notion of linearity – past, present, and future. However, this linear view is but a construct of human cognition. In reality, time may not unfold as neatly as our minds perceive.

Consider a moment of déjà vu, where one feels as though they've experienced the present moment before. Such experiences hint at the non-linear nature of time, suggesting that past, present, and future might be more intricately intertwined than we realize.

Einstein's Theory of Relativity

Albert Einstein revolutionized our understanding of time with his theory of relativity. According to Einstein, time is not absolute; it is relative and can be affected by factors such as gravity and velocity.

In essence, time is intertwined with space to form the fabric of spacetime. Gravity, as described by general relativity, warps this fabric, causing time to flow differently in regions of varying gravitational strength. Thus, a clock placed in a stronger gravitational field would tick slower relative to a clock in a weaker field.

Furthermore, as an object approaches the speed of light, time dilation occurs, causing time to pass more slowly for the moving object relative to a stationary observer. This phenomenon has been experimentally verified, confirming the validity of Einstein's revolutionary insights.

FABRIC OF SPACETIME

Quantum Time

In the realm of quantum mechanics, time takes on a different guise. Unlike classical physics, where time flows uniformly, quantum theory introduces the concept of quantum indeterminacy, where events unfold probabilistically rather than deterministically.

The famous Schrödinger's cat experiment illustrates this principle. Until observed, the state of the cat – whether alive or dead – exists in a superposition of both states. Only upon observation does the wave function collapse, determining the cat's fate.

Similarly, in the quantum realm, time may not adhere to a strict forward progression. Quantum entanglement, where particles become linked regardless of distance, challenges our conventional understanding of causality and temporal sequence.

Time and Consciousness

The relationship between time and consciousness remains a topic of philosophical and scientific debate. Some argue that time is a product of consciousness, a construct that arises from our subjective experience of the world.

Others propose that consciousness itself may transcend time, existing beyond the confines of past, present, and future. Mystics and meditators throughout history have described states of timeless awareness, where the boundaries of individual identity dissolve into a vast, eternal now.

Neuroscientists delve into the workings of the brain, seeking to unravel the neural mechanisms underlying our perception of time. Studies suggest that the brain constructs a sense of temporal order by integrating sensory inputs and memories, creating a coherent narrative of past events.

The Arrow of Time

The concept of the arrow of time describes the asymmetry of time's directionality – the past appears fixed, while the future remains open and uncertain. This arrow is intimately tied to the increase of entropy, or disorder, in the universe.

According to the second law of thermodynamics, entropy tends to increase over time, leading to the inexorable march towards thermodynamic equilibrium. This directional flow of time gives rise to our perception of an irreversible past and an open future.

CHAPTER FOUR

DIMENSIONALITY: EXPLORING THE FABRIC OF REALITY

- Introduction to Dimensionality

In the vast expanse of existence, dimensions serve as the framework upon which the universe is constructed. They define the spatial and temporal coordinates that shape the fabric of reality, providing a means to locate objects, events, and phenomena within the vast tapestry of space-time.

At its core, the concept of dimensionality encompasses the idea of spatial extent—how many directions one can move in a given space. In everyday experience, we are familiar with three spatial dimensions: length, width, and height. These dimensions allow us to navigate and perceive the world around us in three-dimensional space.

However, the notion of dimensionality extends beyond the familiar three dimensions, encompassing a rich and diverse array of mathematical, physical, and philosophical concepts. From the abstract realms of higher-dimensional geometry to the speculative realms of theoretical physics, dimensions play a central role in our understanding of the universe and our place within it.

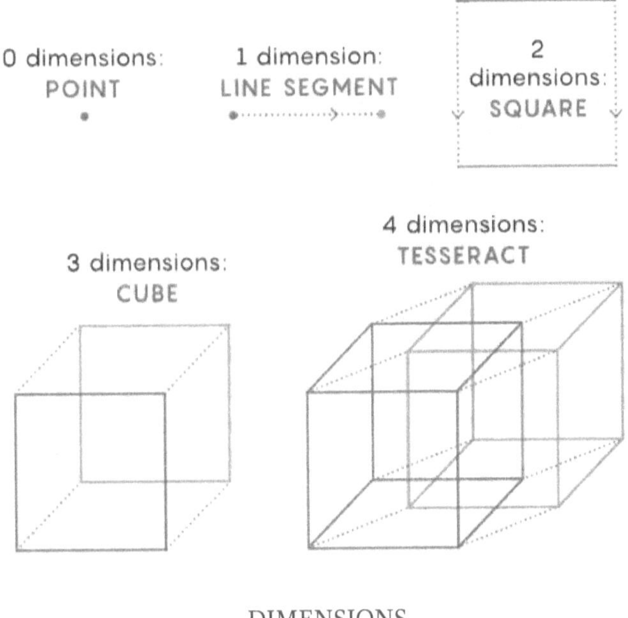

DIMENSIONS

- Mathematical Dimensions

In mathematics, dimensions provide a framework for understanding the geometric properties of space. Euclidean geometry, which forms the foundation of classical geometry, describes space in terms of three dimensions—length, width, and height.

However, mathematics offers a more expansive view of dimensionality, with concepts such as fractal geometry and topological dimensions extending the notion of dimension beyond the confines of Euclidean space.

Fractal geometry, pioneered by mathematicians such as Benoit Mandelbrot, explores structures that exhibit fractional dimensions—neither whole numbers nor integers. Fractals, such as the famous Mandelbrot set, reveal the intricate self-similar patterns that emerge at all scales of magnification, challenging our intuitive understanding of dimensionality.

Topological dimensions, on the other hand, measure the connectivity and complexity of geometric spaces. In topology, a dimension is not necessarily tied to a specific direction but instead reflects the number of independent coordinates needed to specify a point within a space.

- Physical Dimensions

In the realm of physics, dimensions play a fundamental role in describing the structure and behavior of the universe. Albert Einstein's theory of relativity revolutionized our understanding of space and time, presenting a vision of a four-dimensional continuum known as space-time.

According to relativity, space and time are intertwined, forming a unified fabric that is warped and curved by the presence of mass and energy. Gravity, in this framework, arises from the curvature of space-time, with massive objects bending the fabric of reality around them.

Theories such as string theory propose the existence of additional spatial dimensions beyond the familiar three. These extra dimensions, if they exist, may be compactified—curled up into tiny, imperceptible shapes at every point in space. String theory posits that the vibrational modes of tiny strings propagating through these extra dimensions give rise to the diverse array of particles and forces observed in the universe.

- Philosophical Dimensions

Beyond mathematics and physics, dimensions take on a philosophical significance, inviting us to contemplate the nature of reality and our place within it. Philosophers throughout history have grappled with questions of existence, perception, and consciousness, exploring the limits of human understanding.

In philosophy, dimensions serve as a metaphor for the complexity and depth of reality. They represent the multitude of perspectives and interpretations through which we perceive the world around us, from the subjective experience of consciousness to the objective laws of nature.

Questions of existence and identity, explored in philosophical discourses such as metaphysics and ontology, delve into the nature of being and the fundamental structures of reality. Dimensions, in this context, offer a framework for exploring the interconnectedness of all things and the underlying unity that binds the universe together.

- Conclusion: Dimensionality and the Nature of Reality

In conclusion, dimensionality represents a fundamental aspect of the universe, shaping the structure and behavior of reality at every scale. From the mathematical abstractions of higher-dimensional geometry to the empirical discoveries of theoretical physics, dimensions offer a window into the underlying fabric of the cosmos.

As we continue to explore the mysteries of dimensionality, let us embrace the spirit of inquiry and curiosity that drives us to seek understanding of the world around us. For in the exploration of dimensionality lies the potential to unlock the deepest secrets of the universe and uncover the true nature of reality.

CHAPTER FIVE

EXPLORING THE ENIGMA OF BLACK HOLES

In the vast expanse of the universe, amidst the shimmering tapestry of stars and galaxies, lies one of the most enigmatic entities known to humanity: the black hole. Beyond the limits of our comprehension and nestled within the fabric of spacetime, black holes embody the extremes of gravity and density, captivating the imagination of scientists and laypeople alike. Join us on a journey into the heart of darkness as we unravel the mysteries of these cosmic behemoths.

Birth of a Black Hole

Black holes emerge from the cosmic cataclysms that mark the end stages of massive stars. When a star exhausts its nuclear fuel, it undergoes a violent collapse under the relentless pull of gravity. For stars with sufficient mass, this collapse culminates in the formation of a black hole—a point of infinite density known as a singularity, surrounded by an event horizon from which nothing, not even light, can escape.

BLACK HOLE

Anatomy of a Black Hole

Delving deeper into the anatomy of black holes, we explore the intricacies of their structure and properties. From the event horizon, where the fabric of spacetime bends irreversibly, to the inner regions shrouded in darkness, black holes possess a gravitational grip so intense that even time itself is warped and distorted.

Black Hole Varieties

Black holes come in a variety of sizes and flavors, each distinguished by its mass, spin, and electric charge. Stellar-mass black holes, born from the remnants of massive stars, populate the cosmos, while supermassive black holes lurk at the centers of galaxies, exerting their gravitational influence over vast cosmic scales. Intermediate-mass black holes, with masses between those of stellar and supermassive varieties, occupy a more elusive niche in the cosmic hierarchy.

The Dance of Destruction

As cosmic vacuum cleaners, black holes devour anything that ventures too close, from wayward stars to entire clouds of gas and dust. The gravitational forces near a black hole are so extreme that they tear apart any matter unfortunate enough to cross the event horizon, creating a swirling accretion disk of superheated material destined for oblivion.

Unveiling the Secrets

Despite their inscrutable nature, astronomers have devised ingenious methods to study black holes and decipher their mysteries. From observing the effects of gravitational lensing, where the warping of spacetime magnifies distant objects, to detecting the telltale signatures of X-rays and gamma rays emitted by infalling matter, scientists continue to push the boundaries of our understanding.

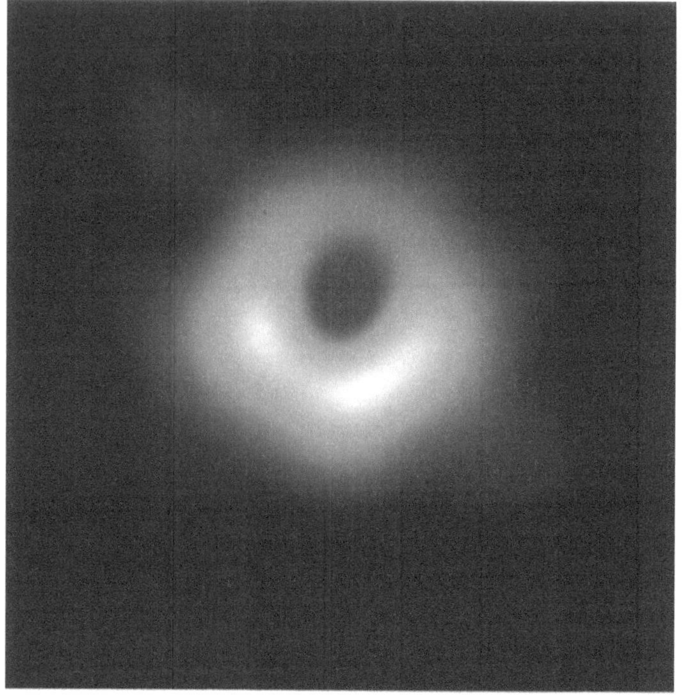

REAL IMAGE OF BLACKHOLE

Black Holes and Time Travel

The extreme gravitational fields surrounding black holes give rise to curious phenomena, including the possibility of time travel. Einstein's theory of relativity predicts that time flows differently in the vicinity of massive objects, leading to the tantalizing prospect of journeying into the past or future by exploiting the gravitational warping of spacetime near a black hole.

Beyond the Event Horizon

What lies beyond the event horizon, the point of no return where the gravitational pull of a black hole becomes insurmountable? Speculations abound, ranging from exotic realms of higher dimensions to the possibility of traversable wormholes connecting distant regions of spacetime. Yet, the true nature of black hole interiors remains a subject of intense debate and speculation.

The Cosmic Role of Black Holes

Beyond their role as cosmic devourers, black holes play a crucial role in shaping the evolution of galaxies and the cosmos at large. Supermassive black holes at the centers of galaxies regulate the growth of their host galaxies through processes such as quasar feedback and galactic winds, influencing the formation of stars and the distribution of matter throughout the universe.

Into the Abyss

As we conclude our journey into the depths of black holes, we are left with a sense of awe and wonder at the cosmic forces that govern the universe. From their humble beginnings as stellar remnants to their towering presence at the centers of galaxies, black holes embody the extremes of nature, challenging our understanding and inspiring new avenues of exploration. Though much remains unknown, the quest to unravel the mysteries of black holes continues, beckoning us ever deeper into the abyss of cosmic discovery.

CHAPTER SIX

Unraveling the Concept of Time Wrapping

- Introduction to Time Wrapping

Time, the inexorable flow that governs the unfolding of events in the universe, has long captivated the human imagination. From the philosophical ponderings of ancient thinkers to the cutting-edge theories of modern physics, the nature of time remains a subject of profound fascination and speculation.

One intriguing concept that arises from our exploration of time is the idea of time wrapping—an idea that challenges our intuitive understanding of temporal progression and suggests the possibility of loops, twists, and folds in the fabric of time itself. But what exactly is time wrapping, and how does it fit into our understanding of the cosmos?

- The Fabric of Spacetime

To understand the concept of time wrapping, we must first delve into the fabric of spacetime—the four-dimensional continuum that unifies space and time, as described by Albert Einstein's theory of relativity. In Einstein's framework, space and time are intertwined, forming a dynamic manifold that is curved by the presence of mass and energy.

According to general relativity, the curvature of spacetime determines the paths that objects follow through the universe, giving rise to the force of gravity. Massive objects such as stars and planets warp the fabric of spacetime around them, creating gravitational wells that dictate the motion of nearby objects.

- Closed Timelike Curves

One of the theoretical constructs that arises from the curvature of spacetime is the notion of closed timelike curves (CTCs). These are hypothetical paths through spacetime that loop back onto themselves, forming closed trajectories that allow an object to return to its own past.

The existence of CTCs raises profound questions about causality and the arrow of time. If time loops are permitted, could they give rise to paradoxes such as the infamous grandfather paradox, where a time traveler could theoretically go back in time and prevent their own existence?

- Wormholes and Time Travel

Theoretical constructs such as wormholes—hypothetical tunnels through spacetime that connect distant regions of the universe—offer a potential mechanism for time wrapping and time travel. According to some theories, traversable wormholes could allow for shortcuts through spacetime, enabling rapid travel between distant points and potentially even facilitating journeys through time.

However, the existence of traversable wormholes remains speculative, and their stability is a subject of ongoing research and debate. The immense energies and exotic forms of matter required to keep a wormhole open raise significant challenges to their practical realization.

WORMHOLES

- Quantum Insights

The intersection of quantum mechanics and general relativity offers further insights into the nature of time wrapping. Quantum phenomena such as entanglement and superposition challenge our classical intuitions about causality and suggest that the fabric of reality may be far more complex and interconnected than previously imagined.

Quantum theories of gravity, such as loop quantum gravity and string theory, seek to reconcile the disparate realms of quantum mechanics and general relativity, offering potential frameworks for understanding the nature of spacetime at the smallest scales.

- Temporal Boundaries

While the concept of time wrapping remains a fascinating theoretical possibility, its practical implications and observational signatures remain elusive. The study of time wrapping pushes the boundaries of our understanding of the universe, inviting us to contemplate the nature of causality, the arrow of time, and the fundamental structure of reality.

As we continue to explore the mysteries of time, from the timeless expanse of the cosmos to the fleeting moments of our everyday lives, let us embrace the spirit of inquiry and curiosity that drives us to seek understanding of the world around us. For in the journey into the depths of time lies the potential to unlock the secrets of the universe and uncover the true nature of existence.

CHAPTER SEVEN

Exploring the Vastness of Time Dilation

In the depths of the universe, where the fabric of spacetime weaves its intricate tapestry, lies a phenomenon that bends our understanding of time itself: time dilation. As humanity delves deeper into the realms of physics and cosmology, the concept of time dilation emerges as one of the most captivating and mind-bending aspects of Einstein's theory of relativity.

1. Unraveling the Concept

Time dilation is a fundamental consequence of Einstein's theory of special relativity, which revolutionized our understanding of space, time, and the relationship between matter and energy. At its core, time dilation proposes that time is not an absolute quantity but rather a variable that can be stretched or contracted depending on the relative motion between observers and the gravitational fields they inhabit.

2. The Twin Paradox: A Thought Experiment

To grasp the essence of time dilation, one must confront the famous Twin Paradox, a thought experiment that illustrates the profound implications of relativistic effects on time. Imagine two identical twins, one embarks on a journey through space at near-light speed while the other remains on Earth. Upon the traveler's return, they find that while only a few years have passed for the spacefarer, decades or even centuries have elapsed for the twin who stayed behind. This stark difference in the passage of time is a direct consequence of time dilation.

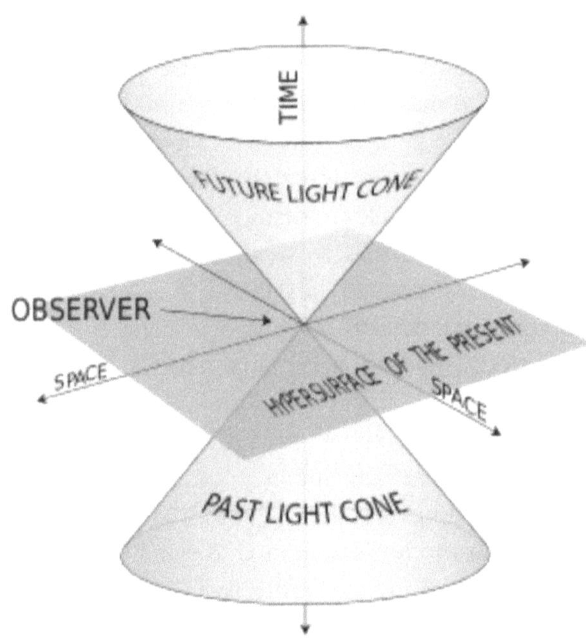

3. Relativity in Action: Experimental Validation

While the Twin Paradox may seem like a mere theoretical conjecture, experimental evidence overwhelmingly supports the reality of time dilation. Experiments involving high-speed particles, such as muons generated in the upper atmosphere and cosmic rays hurtling through space, consistently demonstrate that these particles decay at a slower rate than their stationary counterparts would predict. This observed time dilation confirms the predictions of Einstein's theory and underscores the tangible impact of relativistic effects on the behavior of matter and energy.

4. Spacetime Warping: Gravitational Time Dilation

In addition to velocity-induced time dilation, Einstein's theory of general relativity introduces the concept of gravitational time dilation, where the curvature of spacetime due to mass and energy alters the flow of time itself. According to general relativity, massive objects such as stars and black holes warp the fabric of spacetime, causing time to dilate in their vicinity. This effect manifests as clocks closer to a massive body ticking more slowly relative to clocks situated farther away, a phenomenon known as gravitational time dilation.

5. Practical Applications: Global Positioning System (GPS)

Despite its esoteric nature, time dilation finds practical applications in modern technology, most notably in the Global Positioning System (GPS). The precise synchronization of clocks aboard GPS satellites is crucial for accurate positioning on Earth's surface. However, due to their high orbital speeds and the gravitational potential of Earth, these satellites experience both velocity-induced and gravitational time dilation. Without accounting for these relativistic effects, GPS calculations would quickly drift off course, rendering the system effectively useless. Thus, the successful operation of GPS serves as a real-world validation of the principles of relativity and time dilation.

6. Time Travel: Science Fiction or Plausible Reality?

The notion of time dilation inevitably evokes fantasies of time travel, a staple of science fiction literature and cinema. While time dilation allows for the intriguing possibility of "time dilation" within the framework of relativity, the concept of traversing time freely remains firmly in the realm of speculation. Theoretical constructs such as wormholes and closed timelike curves offer tantalizing glimpses into the theoretical possibility of time travel, but formidable challenges, including the violation of causality and the absence of known physical mechanisms, currently preclude its realization.

7. Beyond the Horizon: Exploring Extreme Regimes

As humanity pushes the boundaries of scientific inquiry, new frontiers in time dilation beckon us to explore extreme regimes where relativistic effects reign supreme. From the vicinity of supermassive black holes, where time slows to a crawl, to the fleeting instants following the Big Bang, where the fabric of spacetime itself undergoes rapid expansion, these cosmic arenas offer tantalizing insights into the nature of time and its intricate relationship with the universe at large.

8. Conclusion: Embracing the Enigma

In the grand tapestry of the cosmos, time dilation stands as a testament to the profound interplay between space, time, and matter. From the mundane to the majestic, from the laboratory to the cosmos, the effects of time dilation permeate every facet of our existence, challenging our intuitions and expanding the horizons of human knowledge. As we gaze into the depths of the universe, let us embrace the enigma of time dilation, for within its mysteries lie the keys to unlocking the secrets of the cosmos itself.

CHAPTER EIGHT

THE SPEED OF LIGHT: EXPLORING THE COSMIC SPEED LIMIT AND ITS IMPLICATIONS FOR TRAVEL

- Introduction to the Speed of Light

The speed of light, denoted by the symbol "c," stands as one of the fundamental constants of the universe. In the realm of physics, it serves as a cosmic speed limit, defining the maximum velocity at which information, energy, and matter can travel through space.

This chapter will explore the significance of the speed of light, its role in shaping our understanding of the cosmos, and the implications it holds for the possibility of interstellar travel.

SPEED OF LIGHT(299,792,458 metres per second)

- The Nature of Light

Light, in its most basic form, consists of electromagnetic waves that propagate through space. As a wave, light exhibits properties such as wavelength, frequency, and amplitude, which determine its characteristics, including color and energy.

One of the most remarkable features of light is its constant speed in a vacuum, approximately 299,792,458 meters per second (or about 186,282 miles per second). This universal speed limit, established by Einstein's theory of relativity, has profound implications for the nature of space and time.

- Einstein's Theory of Relativity

Albert Einstein's theory of relativity, published in the early 20th century, transformed our understanding of space, time, and gravity. Special relativity, introduced in 1905, describes the behavior of objects moving at constant velocities, including the famous equation $E=mc^2$, which equates energy with mass.

General relativity, developed later by Einstein, extends the principles of special relativity to include the effects of gravity. According to general relativity, massive objects warp the fabric of spacetime, influencing the paths of other objects moving through the universe.

- Cosmic Speed Limit

In the framework of relativity, the speed of light emerges as a fundamental constraint on the motion of objects through space. According to special relativity, no object with mass can travel at or exceed the speed of light in a vacuum.

This limitation has far-reaching consequences for the feasibility of interstellar travel, as it imposes severe constraints on the energy requirements and propulsion systems needed to achieve velocities approaching the speed of light.

- Time Dilation and Length Contraction

One of the most intriguing consequences of special relativity is the phenomenon of time dilation, where time appears to slow down for objects moving at relativistic speeds relative to an observer. As an object accelerates toward the speed of light, time dilation becomes increasingly pronounced, leading to apparent discrepancies in the passage of time between observers.

Length contraction is another consequence of special relativity, whereby objects moving at relativistic speeds appear to shrink in the direction of motion from the perspective of a stationary observer.

- Challenges of Interstellar Travel

The cosmic speed limit imposed by the speed of light presents significant challenges for the feasibility of interstellar travel. Even with advanced propulsion systems and energy sources, reaching nearby stars within a human lifetime remains a daunting task.

The immense distances between stars, coupled with the energy requirements and technological constraints of traveling near the speed of light, pose formidable obstacles to the realization of interstellar voyages.

- Advantages of Light-Speed Travel

Despite the challenges, traveling at or near the speed of light offers several potential advantages for interstellar exploration. Time dilation effects could allow travelers to experience shorter subjective journey times, even for voyages spanning vast cosmic distances.

Additionally, relativistic effects such as time dilation and length contraction could provide novel opportunities for space exploration and colonization, enabling humans to reach distant destinations within a reasonable timeframe from their perspective.

- Future Prospects and Conclusion

As humanity continues to push the boundaries of scientific knowledge and technological innovation, the dream of interstellar travel remains an enduring aspiration. While the cosmic speed limit imposed by the speed of light presents formidable challenges, it also opens up new avenues for exploration and discovery.

In the coming decades and centuries, advancements in physics, engineering, and space exploration may bring us closer to the realization of interstellar voyages. Whether through novel propulsion technologies, innovative approaches to space travel, or the harnessing of exotic phenomena in the cosmos, the quest to explore the universe at the speed of light will continue to inspire curiosity, imagination, and exploration.

CHAPTER NINE

WHITE HOLES: THE ENIGMATIC COUNTERPARTS TO BLACK HOLES

- Introduction to White Holes

In the cosmic tapestry of the universe, black holes have long held the spotlight as some of the most enigmatic and intriguing phenomena. Yet, lurking in the shadow of their gravitational dominance lies an equally mysterious counterpart: the white hole.

While black holes are infamous for their ability to trap everything, including light, within their event horizons, white holes are theorized to do the opposite, spewing out matter and energy with unstoppable force. But what exactly are white holes, and how do they fit into our understanding of the cosmos?

WHITE HOLE

- Theoretical Foundations

The concept of white holes emerged from the equations of general relativity, Albert Einstein's revolutionary theory of gravity. In the mathematical framework of general relativity, black holes are formed when massive stars collapse under their own gravity, creating a region of spacetime from which nothing can escape—a phenomenon known as an event horizon.

White holes, on the other hand, are theorized to be the inverse of black holes. Instead of swallowing matter and light, white holes are thought to eject matter and energy outward, creating a region of spacetime from which nothing can enter—a reverse event horizon, so to speak.

While black holes are supported by ample observational evidence, the existence of white holes remains purely theoretical. Nevertheless, their mathematical consistency within the framework of general relativity has led scientists to explore their potential implications for the cosmos.

- The Birth of a White Hole

The formation of a white hole remains a subject of speculation, as no observational evidence of their existence has been confirmed to date. However, one proposed scenario involves the collapse of a black hole followed by a bounce—a theoretical process known as a "white hole bounce."

According to this scenario, as a black hole approaches the end of its lifespan, quantum mechanical effects near the singularity may cause it to undergo a bounce instead of collapsing indefinitely. This bounce would result in the formation of a white hole, ejecting matter and energy outward in a powerful burst of radiation.

While the white hole bounce remains a speculative concept, it offers a tantalizing glimpse into the potential fate of black holes and the possibility of white holes existing as their cosmic counterparts.

Chapter 4: Observational Signatures

Despite their theoretical existence, the observational signatures of white holes remain elusive. Unlike black holes, which leave distinctive imprints on their surroundings through the gravitational effects they exert on nearby matter, white holes are predicted to repel matter and energy, creating a region of spacetime devoid of material.

One potential signature of white holes could be the detection of intense bursts of radiation emanating from their vicinity. These bursts, analogous to the powerful jets emitted by active galactic nuclei, could serve as indirect evidence of the presence of a white hole.

Additionally, the gravitational lensing effects of a white hole on light passing through its vicinity could offer further observational clues to their existence. By studying the distorted images of background objects, astronomers may be able to infer the presence of a white hole lurking in the cosmic depths.

- The Cosmic Landscape

In the grand tapestry of the cosmos, white holes represent yet another fascinating manifestation of the intricate dance between gravity and spacetime. If they exist, white holes could play a significant role in shaping the evolution of galaxies, stars, and even entire cosmic structures.

From their theoretical origins in the equations of general relativity to their speculative implications for the nature of black holes and the fate of the universe, white holes continue to captivate the imagination of scientists and enthusiasts alike. Whether they exist as cosmic anomalies awaiting discovery or as purely mathematical curiosities, their enigmatic presence reminds us of the boundless mysteries that lie beyond the horizon of our current understanding.

CHAPTER TEN

THEORY OF TIME TRAVEL: EXPLORING THE POSSIBILITIES

Introduction

Time travel has long captivated human imagination, inspiring countless stories in literature, film, and scientific speculation. From H.G. Wells' "The Time Machine" to modern movies like "Interstellar," the concept of moving through time, either into the past or the future, raises profound questions about the nature of reality, causality, and the limits of human ingenuity. In this book, we will explore the theoretical underpinnings of time travel, examining how it might be possible according to the laws of physics, and discussing the implications and challenges that such a phenomenon would entail.

- Historical Context and Philosophical Foundations

1. Ancient Myths and Legends

Many ancient cultures have tales of time travel, often involving gods or supernatural forces. These myths reflect humanity's deep fascination with time and our desire to transcend its limitations.

1.2 Philosophical Questions

Time travel raises fundamental philosophical questions. What is time? Is the future predetermined, or do we have free will? Can we change the past? These questions have been debated for centuries by philosophers and scientists alike.

- Understanding Time

2.1 The Nature of Time

Time is often perceived as a linear progression from past to present to future. However, theories such as Einstein's theory of relativity suggest that time is more complex and intertwined with the fabric of space itself.

2.2 Time in Classical Physics

In classical Newtonian mechanics, time is absolute and universal, ticking uniformly for all observers. This view, however, is challenged by modern physics.

2.3 Time in Modern Physics

Einstein's theory of relativity revolutionized our understanding of time. According to relativity, time is relative and can be affected by speed and gravity. This leads to phenomena such as time dilation, where time can pass at different rates for different observers.

- Theoretical Frameworks for Time Travel

3.1 Special Relativity

Special relativity introduces the concept of time dilation, which suggests that at near-light speeds, time slows down for the traveler relative to someone at rest. This effect, although not time travel in the traditional sense, shows that time is not a fixed constant.

3.2 General Relativity

General relativity provides a framework for understanding how gravity can warp space and time. This warping creates the possibility of wormholes, which are theoretical passages through spacetime that could connect distant points in space and time.

3.3 Wormholes and Einstein-Rosen Bridges

A wormhole, or Einstein-Rosen bridge, is a hypothetical tunnel through spacetime. If such structures exist and can be stabilized, they might allow for travel between different times and places in the universe.

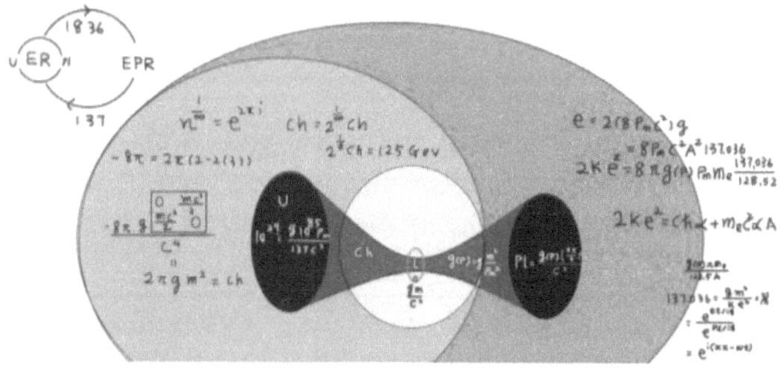

Wormholes and Einstein-Rosen Bridges

- Paradoxes and Challenges

4.1 The Grandfather Paradox

One of the most famous time travel paradoxes is the grandfather paradox, which questions what would happen if a time traveler went back in time and kill their grandfather potentially preventing the time traveler's own existence.

4.2 Causal Loops

Causal loops, or closed time-like curves, are situations where a future event causes a past event, which in turn causes the future event. These loops challenge our understanding of causality and raise questions about the consistency of events in time.

4.3 Resolving Paradoxes

Several theories propose ways to resolve time travel paradoxes, including the many-worlds interpretation, which suggests that each action creates a new parallel universe, and the Novikov self-consistency principle, which asserts that events are consistent with each other, preventing paradoxes.

- Technological and Practical Considerations

5.1 Energy Requirements

The energy required for time travel, particularly for creating and maintaining a wormhole, would be immense, likely far beyond our current technological capabilities.

5.2 Exotic Matter

Stabilizing a wormhole might require exotic matter with negative energy density. Such matter has not been observed and remains hypothetical.

5.3 Human Factors

Time travel poses significant challenges for human travelers, including potential physical and psychological effects of moving through time and adjusting to different temporal environments.

- Time Travel in Science Fiction

6.1 Influential Works

Science fiction has explored time travel in myriad ways, often serving as a mirror for contemporary scientific and philosophical debates. We will review some of the most influential works and how they have shaped public perception of time travel.

6.2 Scientific Accuracy

We will assess the scientific accuracy of various time travel scenarios depicted in fiction, distinguishing between imaginative storytelling and plausible science.

- Ethical and Societal Implications

7.1 Ethical Dilemmas

Time travel introduces ethical dilemmas, such as the consequences of altering past events and the potential exploitation of time travel for personal or political gain.

7.2 Impact on Society

The ability to travel through time could profoundly impact society, affecting everything from our understanding of history to the structure of our legal and economic systems.

7.3 Preventing Misuse

Ensuring that time travel, if possible, is used responsibly would be a significant challenge, requiring robust ethical guidelines and international cooperation.

- Future Prospects and Conclusion

8.1 Current Research

We will review current research in theoretical physics and related fields that may one day lead to practical time travel, highlighting promising avenues and key challenges.

8.2 Speculative Futures

Looking forward, we will speculate on what the future might hold for time travel, considering both optimistic and cautionary scenarios.

8.3 Final Thoughts

Time travel remains one of the most intriguing possibilities in science and philosophy. While significant challenges remain, the pursuit of understanding time and exploring its boundaries continues to drive human curiosity and innovation.

CHAPTER ELEVEN

PART 2 OF TIME TRAVEL

Historical Context and Philosophical Foundations (Continued)

1.3 Early Scientific Speculations

In addition to ancient myths, early scientific thinkers also pondered the nature of time and the possibility of time travel. Philosophers like René Descartes and Isaac Newton laid the groundwork for understanding time as a dimension similar to space, though it would take centuries for these ideas to evolve into the sophisticated theories of today.

1.4 Literary Contributions

Literature has played a crucial role in shaping our conception of time travel. From Mark Twain's "A Connecticut Yankee in King Arthur's Court" to Kurt Vonnegut's "Slaughterhouse-Five," authors have used time travel as a narrative device to explore complex themes of fate, identity, and morality.

Understanding Time (Continued)

2.4 Time and Thermodynamics

The second law of thermodynamics, which states that entropy in a closed system always increases, provides a directionality to time—often referred to as the "arrow of time." This concept helps distinguish between the past and the future, adding another layer of complexity to our understanding of time.

2.5 Quantum Mechanics and Time

Quantum mechanics introduces uncertainty and probability into the fabric of reality. Concepts such as quantum entanglement and superposition challenge classical notions of time, suggesting that at a fundamental level, time may not be as straightforward as it appears.

Theoretical Frameworks for Time Travel (Continued)

3.4 Quantum Tunneling

Quantum tunneling is a phenomenon where particles move through barriers in ways that classical physics cannot explain. Some theorists speculate that similar principles might allow for macroscopic time travel, although this remains highly speculative.

3.5 The Many-Worlds Interpretation

The many-worlds interpretation of quantum mechanics posits that all possible outcomes of a quantum event actually occur, each in a separate, parallel universe. This theory provides a framework for understanding time travel without paradoxes, as traveling to the past would simply create a new timeline.

Paradoxes and Challenges (Continued)

4.4 Information Paradox

The information paradox questions what happens to information that enters a black hole. Some theories suggest that if time travel involves black holes or similar phenomena, resolving the information paradox will be crucial to understanding the mechanics of time travel.

4.5 Temporal Protection Conjecture

Stephen Hawking's temporal protection conjecture proposes that the laws of physics prevent time travel on macroscopic scales, thus averting paradoxes. This conjecture remains a topic of debate and investigation among physicists.

Technological and Practical Considerations (Continued)

5.4 Chronology Protection Conjecture

Similar to Hawking's conjecture, the chronology protection conjecture suggests that the universe inherently forbids time travel to prevent paradoxes. Understanding this conjecture requires advanced knowledge of quantum gravity and the true nature of spacetime.

5.5 Engineering Challenges

Beyond energy requirements and exotic matter, practical time travel would involve engineering challenges such as creating and maintaining stable wormholes, shielding travelers from extreme conditions, and developing navigation systems for temporal journeys.

Time Travel in Science Fiction (Continued)

6.3 Cultural Impact

Science fiction has not only entertained but also influenced scientific research and public attitudes toward time travel. Works like "Doctor Who" and "Back to the Future" have inspired generations of scientists and enthusiasts to explore the boundaries of what might be possible.

6.4 Diverse Perspectives

Time travel in fiction often reflects diverse cultural perspectives. Exploring how different cultures envision and narrate time travel can provide insights into the universal human fascination with time and destiny.

Ethical and Societal Implications (Continued)

7.4 Legal Implications

Time travel would necessitate new legal frameworks to address issues such as crimes committed across different times, rights of individuals in different eras, and the ownership of time-based technologies.

7.5 Economic Impact

The ability to travel through time could revolutionize economies, allowing unprecedented opportunities for investment and resource management but also risking severe economic disparities and exploitation.

Future Prospects and Conclusion (Continued)

8.4 Interdisciplinary Research

Advancements in time travel research will likely require interdisciplinary collaboration, integrating insights from physics, engineering, philosophy, and other fields to address the multifaceted challenges and opportunities.

8.5 Preparing for the Future

As we advance toward the possibility of time travel, preparing for its ethical, social, and technological impacts will be crucial. This preparation involves not only scientific research but also public education and international dialogue.

8.6 Final Thoughts

Time travel remains a tantalizing yet elusive concept. Whether or not it becomes a reality, the quest to understand time will continue to push the boundaries of human knowledge and imagination, driving progress in ways we can scarcely predict. The exploration of time travel is as much a journey of the mind as it is of science, challenging us to rethink our place in the universe.

CHAPTER TWELVE

REAL CASES OF TIME TRAVEL

The idea of real historical time travelers often arises from anecdotal accounts, stories, or supposed evidence that suggest individuals may have somehow journeyed through time. However, these instances are often disputed or debunked, remaining speculative rather than confirmed. Here are a few examples that have captured public interest:

1. **John Titor**: Perhaps one of the most famous modern claims of time travel is attributed to John Titor, who appeared on internet forums in the early 2000s claiming to be a time traveler from the year 2036. Titor shared detailed predictions about future events, technology, and societal changes. However, there's no concrete evidence to support Titor's claims, and many consider it a well-crafted hoax.

2. **The Philadelphia Experiment**: This alleged naval experiment, said to have taken place in 1943 during World War II, supposedly involved attempts to render a ship invisible using electromagnetic fields. Some accounts claim that the experiment inadvertently caused time travel or teleportation effects. However, historical research has

failed to find credible evidence supporting the existence of such an experiment.

3. **Andrew Carlssin**: In 2003, reports surfaced claiming that a man named Andrew Carlssin had been arrested for insider trading and claimed to be a time traveler from the year 2256. The story gained widespread attention, but it was soon revealed to be a hoax perpetrated by a tabloid publication.

4. **Rudolph Fentz**: This tale involves the discovery of a man's body in New York's Times Square in 1950. The man, identified as Rudolph Fentz, allegedly wore clothing from the late 19th century and carried items that seemed out of place for the time. The story suggests that Fentz was a time traveler who accidentally found himself in 1950. However, research has shown that the story originated as a work of fiction by author Jack Finney in a 1951 short story titled "I'm Scared."

5. **The Moberly–Jourdain Incident**: In 1901, two women, Charlotte Anne Moberly and Eleanor Jourdain, claimed to have experienced a time slip while visiting the Palace of Versailles in France. They reported encountering people and scenes from the late 18th century, including Marie Antoinette, despite being in the early 20th century. While their accounts were detailed, they have been widely criticized and are often attributed to psychological factors or creative embellishment.

6. **The Time-Traveling Hipster**: A photograph taken in 1941 during the reopening of the South Forks Bridge in Gold Bridge, Canada, sparked speculation when a man in the crowd appeared to be dressed in modern attire, including sunglasses and a logo-emblazoned T-shirt. Some have suggested that this man might be a time traveler, but closer examination suggests that the clothing and

accessories could be explained by styles of the time.

7. **The Versailles Time Slip**: In 2016, a man claiming to be from the year 2025 posted a series of videos on YouTube describing his alleged experiences of time travel. He claimed to have visited the Palace of Versailles in 2120 and filmed video footage of his journey. However, the authenticity of the videos and the credibility of the claims remain highly dubious.

8. **Flight into the Future**: In 1935, Sir Victor Goddard, a British Royal Air Force officer, claimed to have experienced a time slip while flying over an abandoned airfield near Drem, Scotland. He reported suddenly finding himself flying through a storm and witnessing the restored airfield and planes, as well as people dressed in unfamiliar clothing. Upon returning to base, he discovered that the airfield had been closed and abandoned for years. While intriguing, the story lacks substantial evidence beyond Goddard's own account.

9. **The Time-Traveling Man in Charlie Chaplin's "The Circus"**: In a DVD extra for Charlie Chaplin's 1928 film "The Circus," footage surfaced showing a woman walking by in the background holding what appears to be a mobile phone to her ear. Some viewers speculated that this woman might be a time traveler, though closer examination suggests that she may have been using a portable hearing aid of the time.

10. **The Time-Traveling Watch**: In 2008, Chinese archaeologists claimed to have found a small, intricate metallic watch-like object embedded in a sealed tomb that dated back to the Ming Dynasty (1368-1644). The object, resembling a wristwatch, sparked speculation about time travel, but experts dismissed it as a modern artifact placed in the tomb as a hoax or contamination.

11. **The Case of Rudolph Fenz, the Time-Traveling Taxicab Passenger**: Similar to the Rudolph Fentz tale, another urban legend tells of a man named Rudolph Fenz, who supposedly hailed a taxi in Manhattan in 1950 but vanished during the journey. The driver claimed that Fenz had asked to be taken to a location that no longer existed, leading some to speculate that Fenz was a time traveler who accidentally ended up in 1950 New York.

12. **The Time-Traveling Ancient Greek Laptop**: In 2014, a series of images circulated online showing what appeared to be an ancient Greek sculpture of a woman holding what looked like a laptop computer. The images sparked wild theories about time travel or ancient technology, but they were quickly debunked as modern alterations of the original sculpture.

CHAPTER THIRTEEN

THE CASE OF MIKE MARCUM

Mike "Mad Man" Marcum is an amateur inventor from Stanbury, Missouri who lacks formal science training but has a natural aptitude for electronics. His current project is a modified version of a Jacob's Ladder, a device that consists of two metal rods that start close together at the bottom and spread apart as they go up, producing climbing arcs of electricity. Marcum's idea for a modified version of the Jacob's Ladder is to create a machine that can travel through time. He powered up the machine and produced a small circular area of distortion, or vortex, about 8 inches across that looked like ripples above a fire. It's still unclear what the vortex was or if it was dangerous, so Marcum tossed a sheet metal screw through the field and it vanished. Marcum was stumped and powered down the machine, thinking he had invented some kind of teleportation device. However, he soon realized that the machine was actually a time machine, which he took two weeks off work to test. In this section of the video, we learn about Mike Marcum's invention and how he constructed it from scratch. The invention was a time machine, which used two coils of wire and a climber rod to create an electrical arc and thus travel through time. Marcum created the power supply for the climber rods by coiling wire around a common core and increasing the voltage to 20,000 volts. However, this supply wasn't enough, and Marcum needed more power to test and scale up his invention. He needed access to Transformers that could handle higher voltages than the standard household voltage of 120 or 240 Volts. Marcum needed at least one transformer that could handle 50,000 volts. He needed more than one transformer to be able to scale up the machine properly.

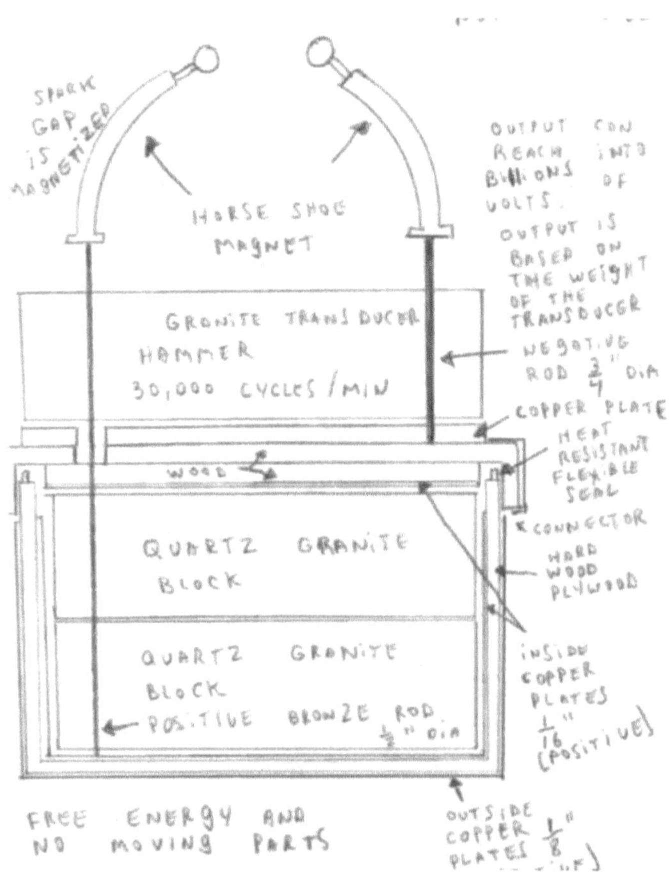

REAL BLUE PRINT OF THE MACHINE MADE BY
MIKE

Mike "Mad Man" Marcum, an amateur inventor from Stanbury, Missouri. He lacks formal science training but has a natural aptitude for electronics. His current project is a modified version of a Jacob's Ladder, a device that consists of two metal rods that start close together at the bottom and spread apart as they go up, producing climbing arcs of electricity. Mike's idea for a modified version of the Jacob's Ladder is to create a machine that can travel through time. He powered up the machine and produced a small circular area of distortion, or vortex, about 8 inches across that looked like ripples above a fire. Mike didn't know what the vortex was or if it was dangerous, so he tossed a sheet metal screw through the field and it vanished. Mike was stumped and powered down the machine, thinking he had invented some kind of teleportation device. However, he soon realized that the machine was actually a time machine, which he took two weeks off work to test.

The invention was a time machine, which used two coils of wire and a climber rod to create an electrical arc and thus travel through time. Marcum created the power supply for the climber rods by coiling wire around a common core and increasing the voltage to 20,000 volts. However, this supply wasn't enough, and Marcum needed more power to test and scale up his invention. He needed access to Transformers that could handle higher voltages than the standard household voltage of 120 or 240 Volts. Marcum needed at least one transformer that could handle 50,000 volts. He needed more than one transformer to be able to scale up the machine properly.

Mike Marcum tells the story of his attempt to build a time machine using six industrial-grade Transformers he stole from a Missouri power company. He loaded the Transformers onto his pickup truck and connected them to the grid, creating an energy field that could send objects through time. However, when he tried to send his cat through the vortex, his cat disappeared. Mike was arrested for stealing the Transformers and causing power outages, and he went to jail for 60 days and five years of probation.

Marcum's time machine. They mention that Mike had considerable knowledge and expertise in electronics, knew about the relationship between turns and voltage, and was able to control duty cycles effectively. Mike came across as humble and not seeking fame or attention, and many people offered to support him in building a bigger version of his machine. collaborative effort to create a new version of the machine, which used rotating magnets. They also mention that the machine was powered by 3 million volts and had access to more power than what was used in the household.

The machine has two cylinders, one inside the other, and uses a magnetic field to rotate a vortex filled with plasma. The speaker mentions The Philadelphia Experiment and how Mike's technology is considered to be on the right track to building a time machine. The machine was tested by sending small objects and animals through the vortex, and Mike predicted where they would end up. The final test involves Mike jumping into the vortex, but when he emerges, he has no memory of his identity or how he got there. The machine, videos, notes, and documentation were all gone from Mike's warehouse when he returned.

Mike "Mad Man" Marcum's third appearance on Coast to Coast, in 2015, found him looking to continue his research and experimentation with time travel. He experienced memory gaps, which brought up issues with fundraising. Mike still couldn't remember donors, supporters or issues surrounding them because his list was in a warehouse. Art Bell encouraged him to keep working, and they discussed crowdfunding, book deals and cooperation with DARPA. Mike even built a tube that could travel through the portal undamaged, a 1930 newspaper article describing a man named John Doe found with a small rectangular device inspired Mike to consider going through the portal and taking his cell phone, notes, and money. Ultimately, no one heard from Mike again, but Art Bell kept tabs on him., stating that he never sent a cat through the portal and that the story of a man washing up on the beach in a metal drum is not true. The speaker then delves into the science behind Mike's time machine, explaining that it would require extremely high gravity, which can be achieved through the use of electromagnetism. It is emphasized that the Earth is always moving in space and that time travel stories often fail to take this into account. The speaker concludes their discussion by reflecting on the importance of Mike's story, stating that he is a true madman who has made remarkable contributions to the field of technology.

CHAPTER FOURTEEN

WORMHOLES

Think about it, if today we use our powerful telescopes to find another planet like Earth in a galaxy, which is suitable for humans to live on, it will take centuries for us to reach it. In fact, for all individuals leaving this galaxy is impossible. Now, everyone knows that the Earth is in the Milky Way Galaxy. And the closest galaxy to the Milky Way is Andromeda Galaxy. Approximately 2.5 million light years away from the Earth. So, if we use a spacecraft to reach there. With the usual speed of 28,000 km per hour it will take 94.5 Billion years to reach it. Not only that, if we can somehow make the technology travel at the speed of light, it will still take 2.5 million years to reach there. This is truly disappointing. What is the point of finding all these planets when we will never be able to travel there? But if there were a shortcut to travel outside the galaxy,a shortcut through which we can travel across million light-years in a few months, then these things become interesting. These shortcuts are Wormholes.

Unraveling the Fabric of Spacetime

At the heart of wormhole theory lies the fabric of spacetime itself, a conceptual framework where space and time intertwine to form a unified continuum. Wormholes, also known as Einstein-Rosen bridges, emerge from the equations of general relativity, the cornerstone of modern physics. According to Einstein's theory, massive objects warp the fabric of spacetime, creating gravitational wells that dictate the motion of celestial bodies. Wormholes represent extreme distortions of spacetime, where the fabric is bent to such an extent that it forms a tunnel-like structure connecting two separate points in space and time.

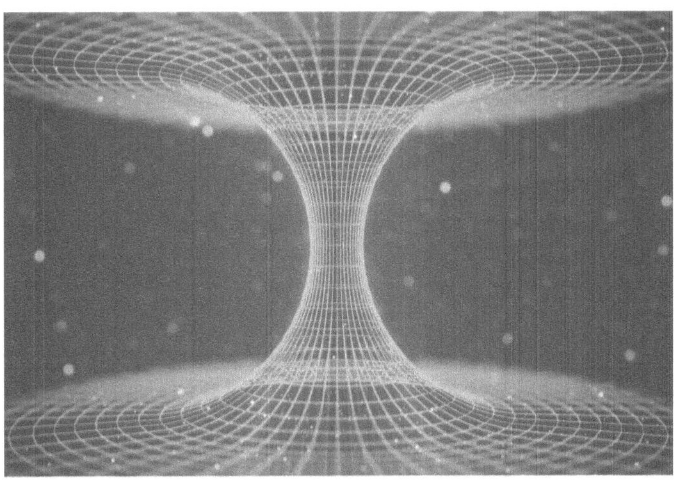

WORMHOLE

Traversing the Cosmic Shortcut

One of the most intriguing aspects of wormholes is their potential as cosmic shortcuts, offering a means to traverse vast cosmic distances in the blink of an eye. In theory, a spacecraft entering one end of a traversable wormhole could emerge at the other end millions of light-years away, effectively circumventing the constraints of conventional space travel. This concept has captured the imagination of scientists and science fiction enthusiasts alike, inspiring visions of interstellar exploration and exotic adventures across the cosmos.

Navigating the Cosmic Minefield

However, the journey through a wormhole is not without its perils. Theoretical models suggest that traversable wormholes may require the existence of exotic matter with negative energy densities to stabilize their structures and prevent their collapse. Moreover, the extreme gravitational forces near the mouths of wormholes could subject travelers to intense tidal forces, potentially tearing apart anything that ventures too close. Navigating these cosmic minefields would require a thorough understanding of the physics governing wormholes and the development of advanced technologies to safely traverse these cosmic conduits.

Unlocking the Secrets of the Universe

Despite the myriad challenges, the study of wormholes offers a glimpse into the deeper mysteries of the universe. By unraveling the nature of these cosmic portals, scientists hope to gain insights into the fundamental workings of spacetime and the nature of gravity itself. Moreover, the existence of traversable wormholes could revolutionize our approach to space exploration, opening up new frontiers in our quest to unravel the mysteries of the cosmos.

Conclusion

As we peer into the cosmic abyss, the allure of wormholes beckons us to embark on a journey of discovery and exploration. While the road ahead is fraught with challenges and uncertainties, the tantalizing promise of unlocking the secrets of the universe spurs us onward, driving humanity ever closer to the realization of our cosmic destiny. In the timeless dance of space and time, wormholes stand as portals to new realms, inviting us to venture boldly into the unknown and chart a course towards the stars.

Summary

Wormholes, a concept based on real science, could potentially provide a shortcut for intergalactic travel. While their existence has not been proven, wormholes are a solution to Einstein's field equations. White holes, the theoretical opposite of black holes, could potentially serve as the exit point for a wormhole. However, the existence of white holes is still debated among scientists.

Highlights

? Wormholes are a potential shortcut for intergalactic travel, allowing us to traverse millions of light-years in a matter of minutes.

? The concept of wormholes is based on real science and is a solution to Einstein's field equations.

? Wormholes could potentially be formed by the extreme gravitational force of black holes, with white holes serving as their exit points.

? While wormholes and white holes are currently theoretical concepts, the discovery and study of black holes were also initially theoretical but were later proven.

? Scientists have successfully created a magnetic wormhole in a laboratory, demonstrating the possibility of wormhole-like phenomena in magnetic fields.

? Other theoretical solutions for faster-than-light travel, such as the Alcubierre Drive, have been proposed but require further exploration and development.

? The existence and practicality of wormholes and other methods of faster-than-light travel are still subjects of scientific debate and research.

Key Insights

? The concept of wormholes is based on the bending and curvature of space-time, as described by Einstein's field equations. These equations explain how matter and energy influence the fabric of space-time.

? Wormholes offer the potential for intergalactic travel by providing a shortcut through the fabric of space-time. If they exist, they could allow us to travel vast distances in a fraction of the time it would take through conventional means.

? White holes, the theoretical opposite of black holes, could potentially serve as the exit points for wormholes. These hypothetical objects would emit light and matter instead of absorbing them, allowing for passage from one region of space-time to another.

? While the existence of wormholes and white holes is still theoretical, the discovery and study of black holes provide hope that these phenomena could be observed and understood in the future.

? Scientists have already made progress in creating wormhole-like phenomena in magnetic fields, demonstrating that such concepts are not purely science fiction.

? The practicality and feasibility of wormholes and other methods of faster-than-light travel, such as the Alcubierre Drive, are still uncertain and require further scientific research and exploration.

? The study of wormholes and other potential shortcuts for intergalactic travel opens up exciting possibilities for the future of space exploration and our understanding of the universe.

CHAPTER FIFTEEN

THE TWIN PARADOX: A JOURNEY THROUGH RELATIVISTIC TIME

In the annals of physics, few thought experiments have captured the imagination of scientists and laypersons alike as profoundly as the Twin Paradox. This paradox, rooted in Einstein's theory of special relativity, presents a scenario that challenges our intuitive understanding of time and space, offering profound insights into the nature of reality itself.

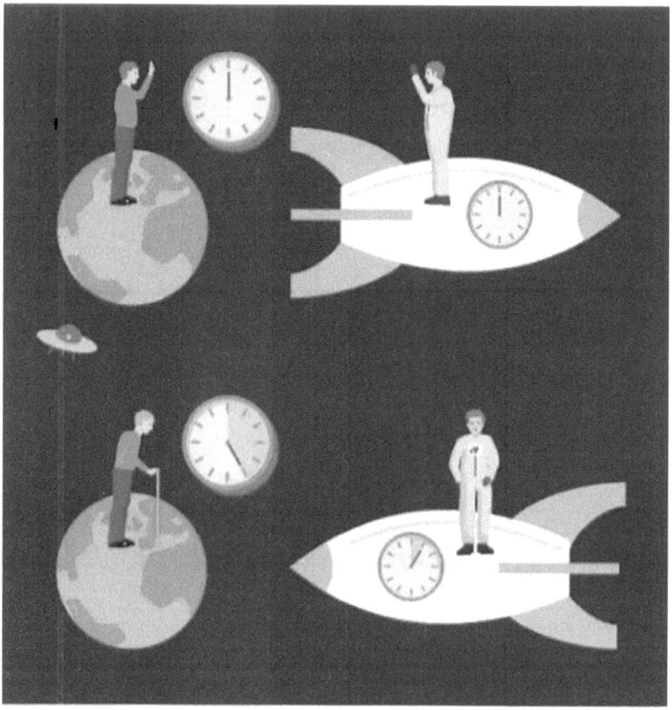

TWIN PARADOX

1. Setting the Stage: The Thought Experiment Unfolds
Imagine two identical twins, born on the same day and sharing an unbreakable bond. One twin, let's call her Alice, remains on Earth, while the other, named Bob, embarks on a cosmic journey through space. Bob's mission is ambitious: to explore the cosmos at velocities nearing the speed of light, pushing the boundaries of human exploration.

2. The Journey Begins: Relativistic Effects in Motion

As Bob's spacecraft accelerates to relativistic speeds, strange and wondrous phenomena begin to unfold. According to Einstein's theory of special relativity, as an object accelerates closer to the speed of light, time for that object appears to slow down relative to observers at rest. From Bob's perspective, time aboard his spacecraft appears to flow normally. However, for Alice, observing Bob's journey from the confines of Earth, time aboard the spacecraft appears to slow down significantly.

3. Reunion Deferred: The Relativity of Simultaneity

As Bob hurtles through the cosmos, decades pass for Alice back on Earth. From her perspective, Bob's aging process slows down, and he appears to age at a much slower rate than herself. This apparent contradiction arises from the relativity of simultaneity – the idea that events that appear simultaneous to one observer may not be simultaneous to another, depending on their relative motion.

4. The Return Journey: Resolving the Paradox

Eventually, Bob's cosmic voyage comes to an end, and he returns to Earth to reunite with his twin sister, Alice. To his surprise, while he may have experienced only a few years aboard his spacecraft, Alice has aged several decades in his absence. The paradoxical nature of this scenario arises from the fact that Bob's journey involved acceleration and deceleration, breaking the symmetry between the two twins. This acceleration introduces a preferred frame of reference, resolving the apparent paradox and reconciling the differing experiences of time between the two twins.

5. Real-World Implications: Practical Applications and Experimental Evidence

While the Twin Paradox may seem like a purely theoretical curiosity, its implications extend far beyond the realm of thought experiments. Experimental evidence supporting the predictions of special relativity abounds, from high-energy particle accelerators to cosmic ray observations. Indeed, the phenomenon of time dilation has practical implications in various fields, from space exploration to the operation of global positioning systems.

6. Philosophical Reflections: Time, Reality, and Perception

The Twin Paradox invites us to ponder the nature of time itself – is it an immutable river flowing inexorably forward, or a malleable construct shaped by the fabric of spacetime? Moreover, the paradox prompts us to question the nature of reality and our perception thereof. Are our experiences of time truly objective, or are they contingent upon our relative motion and frame of reference?

7. Conclusion: A Window into the Relativistic Universe

In the grand tapestry of the cosmos, the Twin Paradox stands as a shining example of the profound insights afforded by Einstein's theory of relativity. From its humble origins as a thought experiment to its real-world implications in the realms of physics and philosophy, the Twin Paradox offers a window into the relativistic universe, where time is not an absolute quantity but a fluid, ever-changing entity shaped by the dynamics of motion and gravity. As we continue to explore the frontiers of science and push the boundaries of human knowledge, let us not forget the lessons imparted by the Twin Paradox – that the universe is far stranger and more wondrous than we can possibly imagine.

CHAPTER SIXTEEN

TIME LOOPS AND BOOTSTRAP PARADOXES: NAVIGATING THE RECURSIVE NATURE OF CAUSALITY

In the labyrinthine corridors of theoretical physics and speculative fiction, few concepts are as beguiling and confounding as time loops and bootstrap paradoxes. These phenomena, rooted in the intricate fabric of time travel and causality, challenge our notions of causation, identity, and the flow of time itself, offering tantalizing glimpses into the paradoxical nature of existence.

1. The Essence of Time Loops: A Recursive Journey Through Time

A time loop, in its simplest form, is a sequence of events that repeats itself indefinitely, forming a closed causal circuit in which the past influences the future and vice versa. Imagine a traveler journeying back in time to a specific moment, only to inadvertently cause the events that lead to their own journey, thus perpetuating the loop ad infinitum. In such scenarios, the boundary between cause and effect becomes blurred, as each iteration of the loop reinforces the conditions for its own existence.

2. Bootstrap Paradoxes: Objects Without Origins

At the heart of many time loops lies the enigmatic Bootstrap Paradox, wherein an object or information is passed from the future to the past, creating a causal loop with no discernible origin. One of the most famous examples of this paradox is the scenario of a composer who receives a masterpiece from the future, only to later travel back in time and present the same composition to their younger self, thus initiating the loop. In such cases, the origin of the object or information becomes indeterminate, leading to profound questions about causality and the nature of existence.

3. Resolutions and Interpretations: Untangling the Knots of Time

Despite their apparent logical conundrums, several resolutions and interpretations of time loops and bootstrap paradoxes have been proposed. Some theorists suggest that the existence of parallel timelines or branching universes may alleviate the paradoxical contradictions, allowing for the coexistence of multiple causally consistent realities. Others argue for the existence of a self-consistent "block universe," where past, present, and future coexist simultaneously, rendering questions of causality moot.

4. Philosophical Reflections: Time, Identity, and the Nature of Reality

Time loops and bootstrap paradoxes invite profound philosophical reflections on the nature of time, identity, and the fabric of reality itself. Do these paradoxes imply a deterministic universe, where the course of history is predetermined and immutable? Or do they suggest a more nuanced understanding of causality, wherein the past, present, and future are inextricably intertwined, shaping and reshaping one another in a continuous dance of existence? These questions lie at the heart of the paradoxes and continue to provoke deep contemplation among scholars and thinkers.

5. Conclusion: Embracing the Paradoxical Nature of Existence

In the grand tapestry of the cosmos, time loops and bootstrap paradoxes stand as poignant reminders of the inherent complexity and mystery of existence. From their humble origins as speculative thought experiments to their profound implications for our understanding of causality and identity, these paradoxes challenge us to confront the inherent contradictions and uncertainties of the universe. As we navigate the recursive landscape of time and causality, let us not shy away from the paradoxes that confront us, but instead, embrace them as portals into the rich and multifaceted nature of reality itself.

CHAPTER SEVENTEEN

EXTRATERRESTRIAL LIFE

Extraterrestrial life, often abbreviated as ET life or simply referred to as aliens, is a topic that has fascinated humanity for centuries. Here's an overview:

Definition: Extraterrestrial life refers to life forms that exist outside of Earth. This could include organisms ranging from simple microbes to complex intelligent beings.

Search for Extraterrestrial Intelligence (SETI): Scientists have been actively searching for signs of extraterrestrial intelligence through projects like SETI, which use radio telescopes to listen for signals from other civilizations. While no conclusive evidence has been found yet, the search continues.

Exoplanets, or extrasolar planets, are planets that orbit stars outside of our solar system. Their discovery has revolutionized our understanding of planetary systems and the prevalence of planets in the universe. Here's a closer look at exoplanets:

Discovery: The first confirmed detection of an exoplanet orbiting a sun-like star was announced in 1995, with the discovery of 51 Pegasi b. Since then, advancements in technology and observation techniques have led to the discovery of thousands of exoplanets.

Detection Methods:

Radial Velocity Method: This technique detects exoplanets by measuring the slight wobble or "wobble" in a star's motion caused by the gravitational pull of orbiting planets.

Transit Method: Exoplanets can also be detected when they pass in front of their host star, causing a temporary decrease in the star's brightness as seen from Earth.

Direct Imaging: Directly capturing images of exoplanets is challenging due to the brightness of their host stars. However, advancements in telescope technology have enabled the direct imaging of some larger exoplanets.

Gravitational Microlensing: This method detects exoplanets by observing the bending of light from distant stars due to the gravitational influence of foreground objects, such as planets.

Types of Exoplanets:

Hot Jupiters: Gas giant planets that orbit very close to their host stars, resulting in high temperatures.

Super-Earths: Rocky planets with masses greater than Earth but less than Neptune.

Mini-Neptunes: Gaseous planets with sizes intermediate between Earth and Neptune.

Terrestrial Planets: Rocky planets similar in size and composition to Earth.

Rogue Planets: Planets that do not orbit any star and instead drift freely through space.

Habitability: Scientists are particularly interested in identifying exoplanets within the "habitable zone" of their host stars, where conditions may be suitable for liquid water to exist on the planet's surface. Such planets are considered potential candidates for hosting life as we know it.

Kepler Space Telescope: Launched by NASA in 2009, the Kepler Space Telescope played a crucial role in the discovery of exoplanets by using the transit method to survey a portion of the Milky Way galaxy.

Follow-up Studies: Follow-up observations of exoplanets, conducted using ground-based telescopes and space-based observatories like the Hubble Space Telescope, provide additional insights into their atmospheres, compositions, and physical characteristics.

The study of exoplanets continues to expand our knowledge of planetary formation, the diversity of planetary systems, and the potential for finding habitable worlds beyond our solar system. Ongoing efforts, including the development of next-generation telescopes and space missions, promise to uncover even more exoplanets and further unravel the mysteries of the cosmos.

Microbial life, often referred to simply as microbes, constitutes a diverse and ubiquitous form of life found throughout the Earth's biosphere. Here's an overview:

Definition: Microbial life encompasses various microscopic organisms, including bacteria, archaea, fungi, protists, and viruses. These organisms are typically too small to be seen with the naked eye and are found in virtually every environment on Earth, from deep-sea vents to polar ice caps.

Role in the Biosphere: Microbes play crucial roles in ecosystems and biogeochemical processes. They are involved in nutrient cycling, decomposition, soil formation, and symbiotic relationships with plants and animals. Microbes also play significant roles in human health, agriculture, industry, and environmental remediation.

Extreme Environments: Microbial life is remarkably adaptable and can thrive in extreme environments that were once thought to be uninhabitable. These include acidic hot springs, deep-sea hydrothermal vents, salt flats, frozen environments, and even within the human body.

Origin of Life: Microbial life is thought to have been the earliest form of life on Earth, appearing billions of years ago. The study of extremophiles—microbes that thrive in extreme environments—provides insights into the conditions that may have existed on early Earth and other planets.

Astrobiology: Astrobiologists study microbial life on Earth to understand the potential for life elsewhere in the universe. Extremophiles are of particular interest in astrobiology, as they offer clues about the types of environments that could support life on other planets or moons.

Microbial Ecology: Microbial ecologists study the interactions between microbes and their environments to understand microbial community dynamics, biodiversity, and ecosystem functioning. Advances in DNA sequencing technologies have revolutionized the field by allowing researchers to study microbial communities in greater detail.

Applied Microbiology: Microbes have numerous practical applications in biotechnology, medicine, agriculture, and environmental science. They are used to produce antibiotics, enzymes, vaccines, biofuels, biodegradable plastics, and other valuable products. Microbes are also used in wastewater treatment, soil restoration, and the bioremediation of contaminated sites.

Challenges and Opportunities: While microbes offer many benefits, they can also pose challenges, such as causing infectious diseases, food spoilage, and environmental degradation. Understanding microbial diversity, evolution, and interactions is essential for addressing these challenges and harnessing the full potential of microbial life.

Overall, microbial life is a fundamental component of Earth's biosphere, with profound implications for the past, present, and future of life on our planet and beyond.

he Fermi Paradox is a thought-provoking concept that highlights the apparent contradiction between the high probability of the existence of extraterrestrial civilizations and the lack of evidence for, or contact with, such civilizations. Here's a deeper dive into the Fermi Paradox:

Origins: The Fermi Paradox is named after physicist Enrico Fermi, who, during a conversation about the possibility of extraterrestrial life in 1950, famously asked, "Where is everybody?" This simple question encapsulates the essence of the paradox.

Probability of Extraterrestrial Life: Given the vast number of stars and planets in the observable universe, scientists estimate that there should be a large number of potentially habitable planets capable of supporting life. The sheer scale of the universe suggests that the probability of extraterrestrial civilizations emerging elsewhere is high.

Lack of Evidence: Despite the high probability of extraterrestrial civilizations, no conclusive evidence of their existence has been found. This absence of evidence stands in stark contrast to the expectation that if advanced civilizations were common, we should have detected their signals, observed their technological artifacts, or even encountered them directly.

Possible Explanations:

Rare Earth Hypothesis: Some theories propose that the conditions necessary for complex life, such as Earth's stable climate, plate tectonics, and large moon, are rare in the universe, limiting the emergence of intelligent civilizations.

Great Filter: The Great Filter hypothesis suggests that there may be significant barriers or challenges that prevent life from advancing beyond a certain stage of development. These barriers could include catastrophic events, such as asteroid impacts, or societal factors, such as self-destruction through warfare or environmental degradation.

Technological Singularity: Some speculate that advanced civilizations may reach a technological singularity, a point at which their technological development accelerates dramatically, leading to the creation of advanced artificial intelligence or the ability to upload consciousness into virtual environments. Such civilizations may become less detectable or interested in traditional forms of communication.

Rare Communication: It's possible that advanced civilizations exist but are simply not broadcasting detectable signals in ways that we can currently detect. They may use alternative forms of communication, such as neutrino beams or advanced encryption methods, or they may intentionally remain hidden to avoid contact with less advanced civilizations.

Interstellar Travel Limitations: Despite the vastness of the universe, the challenges of interstellar travel, such as the vast distances and energy requirements, may make direct contact between civilizations exceedingly rare, even if they exist.

Implications: The Fermi Paradox raises profound questions about humanity's place in the cosmos, the nature of extraterrestrial life, and the future of our own civilization. It challenges scientists, philosophers, and the public to contemplate the possibilities and limitations of our understanding of the universe.

The Fermi Paradox remains an enduring mystery and continues to stimulate scientific inquiry and speculation about the nature of intelligent life in the universe.

Voyager 1

The Voyager missions, particularly Voyager 1 and Voyager 2, have made significant contributions to our understanding of the universe and potential interactions with extraterrestrial life, albeit indirectly.

Golden Record: Perhaps the most famous aspect of the Voyager missions in this context is the inclusion of the Golden Record. Each Voyager spacecraft carries a gold-plated phonograph record containing sounds and images selected to portray the diversity of life and culture on Earth. The idea behind this was to communicate a message from humanity to any intelligent extraterrestrial life that might encounter the spacecraft.

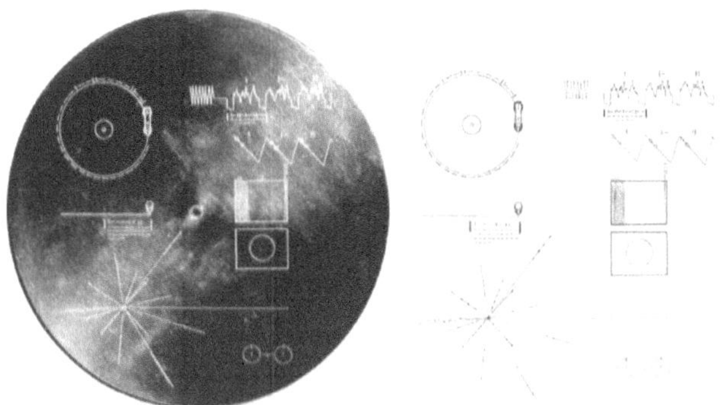

The Golden Record cover shown with its extraterrestrial instructions. Credit: NASA/JPL

THE GOLDEN RECORD

Interstellar Space: Both Voyager 1 and Voyager 2 have exited the heliosphere, the protective bubble of particles and magnetic fields created by the Sun, and entered interstellar space. As they continue to journey through the galaxy, they provide valuable data on the interstellar medium and the conditions beyond our solar system, which could indirectly inform our understanding of potential habitats for extraterrestrial life.

Exploration of Outer Planets: Prior to leaving the solar system, Voyager 1 and Voyager 2 conducted extensive exploration of the outer planets and their moons, including Jupiter, Saturn, Uranus, and Neptune. Their observations provided unprecedented insights into the geology, atmospheres, and magnetospheres of these distant worlds, offering clues about the potential for habitability and the conditions necessary for life elsewhere in the universe.

Continued Communication: While the primary mission of the Voyager spacecraft was to explore the outer solar system, they continue to communicate with Earth even in interstellar space. This ongoing communication allows scientists to study the effects of the interstellar medium on the spacecraft and their instruments, as well as providing a platform for potential future communication with extraterrestrial intelligence, should such encounters occur.

In summary, while the Voyager missions were not specifically designed to directly interact with extraterrestrial life, they have made valuable contributions to our understanding of the cosmos, provided a message from humanity in the form of the Golden Record, and continue to serve as ambassadors of Earth as they journey through interstellar space.

Part 2: Extraterrestrial Life and the Search for Exoplanets Summary

Discovery and Types of Exoplanets

2.1 Discovery of Exoplanets

The discovery of exoplanets—planets that orbit stars outside our solar system—has revolutionized our understanding of the universe. The first confirmed detection of an exoplanet orbiting a sun-like star, 51 Pegasi b, in 1995 marked a significant milestone. Since then, thousands of exoplanets have been discovered, thanks to advances in technology and observational techniques.

2.2 Detection Methods

Radial Velocity Method

This method detects exoplanets by measuring the wobble in a star's motion caused by the gravitational pull of orbiting planets. This wobble results in variations in the star's spectral lines due to the Doppler effect.

Transit Method

Exoplanets can also be detected when they pass in front of their host star, causing a temporary dip in the star's brightness as seen from Earth. This method has been particularly successful, with missions like NASA's Kepler Space Telescope discovering thousands of exoplanets using this technique.

Direct Imaging

Directly capturing images of exoplanets is challenging due to the brightness of their host stars. However, with advancements in telescope technology, some larger exoplanets have been directly imaged, providing valuable insights into their atmospheres and compositions.

Gravitational Microlensing

This method involves observing the bending of light from a distant star due to the gravitational influence of a foreground object, such as a planet. It can detect planets that are otherwise difficult to observe using other methods.

2.3 Types of Exoplanets

Hot Jupiters

Gas giant planets that orbit very close to their host stars, resulting in extremely high temperatures. Their proximity to their stars makes them easier to detect.

Super-Earths

Rocky planets with masses greater than Earth but less than Neptune. These planets are of particular interest because they might possess conditions suitable for life.

Mini-Neptunes

Gaseous planets with sizes between Earth and Neptune. Their atmospheric compositions provide clues about planetary formation and evolution.

Terrestrial Planets

Rocky planets similar in size and composition to Earth. These planets are prime candidates in the search for life because they might have Earth-like conditions.

Rogue Planets

Planets that do not orbit any star and drift freely through space. Their origins and characteristics remain subjects of ongoing research.

2.4 Habitability

Scientists are particularly interested in exoplanets within the "habitable zone" of their host stars, where conditions might allow for liquid water to exist on the planet's surface. Such planets are considered potential candidates for hosting life as we know it.

2.5 Kepler Space Telescope

Launched by NASA in 2009, the Kepler Space Telescope significantly advanced the discovery of exoplanets. By using the transit method to survey a portion of the Milky Way galaxy, Kepler identified thousands of exoplanet candidates, greatly expanding our knowledge of planetary systems.

2.6 Follow-up Studies

Follow-up observations of exoplanets using ground-based telescopes and space-based observatories like the Hubble Space Telescope provide additional insights into their atmospheres, compositions, and physical characteristics. These studies help refine our understanding of exoplanet diversity and habitability.

Microbial Life and Its Implications

3.1 Definition and Role in the Biosphere

Microbial life, which includes bacteria, archaea, fungi, protists, and viruses, constitutes a diverse and ubiquitous form of life on Earth. Microbes play crucial roles in ecosystems and biogeochemical processes, such as nutrient cycling, decomposition, soil formation, and symbiotic relationships with plants and animals.

3.2 Extreme Environments

Microbial life is remarkably adaptable and can thrive in extreme environments once thought uninhabitable. These include acidic hot springs, deep-sea hydrothermal vents, salt flats, frozen environments, and even within the human body. The study of extremophiles—microbes that thrive in extreme conditions—provides insights into the potential for life on other planets or moons with harsh environments.

3.3 Origin of Life and Astrobiology

Microbial life is thought to be the earliest form of life on Earth, appearing billions of years ago. Astrobiologists study these ancient life forms to understand the conditions that may have existed on early Earth and other planets. Extremophiles, in particular, offer clues about the types of environments that could support life beyond our planet.

3.4 Microbial Ecology

Microbial ecologists investigate the interactions between microbes and their environments to understand community dynamics, biodiversity, and ecosystem functioning. Advances in DNA sequencing technologies have revolutionized this field, allowing researchers to study microbial communities in unprecedented detail.

3.5 Applied Microbiology

Microbes have numerous practical applications in biotechnology, medicine, agriculture, and environmental science. They are used to produce antibiotics, enzymes, vaccines, biofuels, biodegradable plastics, and other valuable products. Microbes are also employed in wastewater treatment, soil restoration, and the bioremediation of contaminated sites.

3.6 Challenges and Opportunities

While microbes offer many benefits, they also pose challenges, such as causing infectious diseases, food spoilage, and environmental degradation. Understanding microbial diversity, evolution, and interactions is essential for addressing these challenges and harnessing the full potential of microbial life.

The Fermi Paradox

4.1 Origins and Core Question

The Fermi Paradox, named after physicist Enrico Fermi, highlights the contradiction between the high probability of extraterrestrial civilizations' existence and the lack of evidence for or contact with such civilizations. In 1950, Fermi famously asked, "Where is everybody?" This question encapsulates the paradox's essence.

4.2 Probability of Extraterrestrial Life

Given the vast number of stars and planets in the observable universe, scientists estimate that many potentially habitable planets exist. The sheer scale of the universe suggests that the probability of extraterrestrial civilizations emerging elsewhere is high.

4.3 Lack of Evidence

Despite the high probability of extraterrestrial civilizations, no conclusive evidence of their existence has been found. This absence of evidence contrasts with the expectation that if advanced civilizations were common, we should have detected their signals, observed their technological artifacts, or even encountered them directly.

4.4 Possible Explanations

Rare Earth Hypothesis

Some theories propose that the conditions necessary for complex life, such as Earth's stable climate, plate tectonics, and large moon, are rare in the universe, limiting the emergence of intelligent civilizations.

Great Filter

The Great Filter hypothesis suggests that significant barriers or challenges prevent life from advancing beyond a certain stage of development. These barriers could include catastrophic events, such as asteroid impacts, or societal factors, such as self-destruction through warfare or environmental degradation.

Technological Singularity

Some speculate that advanced civilizations may reach a technological singularity, a point at which their technological development accelerates dramatically, leading to the creation of advanced artificial intelligence or the ability to upload consciousness into virtual environments. Such civilizations may become less detectable or interested in traditional forms of communication.

Rare Communication

It is possible that advanced civilizations exist but do not broadcast detectable signals in ways we can currently detect. They may use alternative forms of communication, such as neutrino beams or advanced encryption methods, or intentionally remain hidden to avoid contact with less advanced civilizations.

Interstellar Travel Limitations

Despite the vastness of the universe, the challenges of interstellar travel, such as the vast distances and energy requirements, may make direct contact between civilizations exceedingly rare, even if they exist.

4.5 Implications

The Fermi Paradox raises profound questions about humanity's place in the cosmos, the nature of extraterrestrial life, and the future of our own civilization. It challenges scientists, philosophers, and the public to contemplate the possibilities and limitations of our understanding of the universe. The Fermi Paradox remains an enduring mystery and continues to stimulate scientific inquiry and speculation about the nature of intelligent life in the universe.

- SUMMARY OF VOYAGER MISSION (from part 1)

The Voyager Missions

5.1 Voyager 1 and 2 Overview

The Voyager missions, particularly Voyager 1 and Voyager 2, have made significant contributions to our understanding of the universe and potential interactions with extraterrestrial life, albeit indirectly.

5.2 The Golden Record

Each Voyager spacecraft carries a gold-plated phonograph record containing sounds and images selected to portray the diversity of life and culture on Earth. This "Golden Record" serves as a message from humanity to any intelligent extraterrestrial life that might encounter the spacecraft. The contents of the Golden Record were curated to provide a broad representation of Earth's biosphere and human culture.

5.3 Interstellar Space

Both Voyager 1 and Voyager 2 have exited the heliosphere, the protective bubble of particles and magnetic fields created by the Sun, and entered interstellar space. As they continue to journey through the

CHAPTER EIGHTEEN

EXPLORING THE ENIGMA OF REALITY: ARE WE LIVING IN A SIMULATION?

n the labyrinth of human thought, there exists a tantalizing conjecture that challenges the very fabric of our understanding of reality: the simulation hypothesis. This provocative idea posits that our universe, with all its complexities and intricacies, may be nothing more than a digital construct, a simulated reality created by beings or entities far more advanced than ourselves. As we embark on this journey of exploration, we delve into the depths of philosophical inquiry, scientific speculation, and existential contemplation, seeking to unravel the mysteries that shroud the nature of existence itself.

1. The Genesis of the Simulation Hypothesis

The roots of the simulation hypothesis can be traced back to ancient philosophical traditions, where thinkers grappled with questions of reality, illusion, and the nature of perception. However, it was not until the advent of modern computing and virtual reality technologies that the idea gained traction in scientific and philosophical discourse. In 2003, philosopher Nick Bostrom reignited interest in the concept with his seminal paper "Are You Living in a Computer Simulation?" In this paper, Bostrom presented a thought-provoking argument that posited the existence of highly advanced civilizations capable of creating simulated realities inhabited by conscious beings.

2. The Plausibility of Simulated Realities

Advancements in computer science, artificial intelligence, and virtual reality have blurred the lines between the virtual and the real, raising questions about the feasibility of creating simulated worlds indistinguishable from our own. Proponents of the simulation hypothesis argue that if civilizations like ours continue to advance technologically, it may be possible to create vast virtual universes populated by simulated beings with complex cognitive faculties and subjective experiences. Moreover, the exponential growth of computational power suggests that the creation of such simulations may become feasible in the not-too-distant future.

3. Philosophical Underpinnings and Metaphysical Implications

At the heart of the simulation hypothesis lie profound philosophical questions about the nature of reality, consciousness, and existence. Concepts such as Cartesian doubt and solipsism – the philosophical skepticism about the external world and the existence of other minds – take on new significance in the context of a simulated reality. Furthermore, the simulation hypothesis challenges traditional notions of free will, morality, and the purpose of existence, inviting us to reconsider our fundamental assumptions about the nature of reality itself.

4. Evidence and Anomalies in the Simulation

While there is currently no direct empirical evidence to confirm or refute the simulation hypothesis, proponents point to potential "glitches" or anomalies in our reality as possible indicators of simulation artifacts. These anomalies could range from unexplained phenomena in quantum mechanics to inconsistencies in the laws of physics that may hint at underlying computational constraints. However, skeptics argue that such anomalies can be explained by known scientific principles and do not necessarily require invoking the existence of a simulated reality.

5. Ethical Considerations and Existential Dilemmas

The prospect of living in a simulated reality raises profound ethical and existential questions about the nature of consciousness, agency, and the human experience. If our reality is indeed a simulation, what implications does this have for our understanding of morality and responsibility? Furthermore, how would the knowledge of our simulated nature impact our sense of identity and existential outlook? These questions challenge us to confront the very essence of what it means to be human and to grapple with the ethical implications of our potential existence within a simulated world.

6. The Search for Truth: Scientific Inquiry and Speculative Endeavors

While the simulation hypothesis remains speculative and unproven, it continues to inspire scientific inquiry and philosophical reflection. Researchers in fields ranging from cosmology to computer science are exploring the theoretical underpinnings of simulated realities and seeking ways to test the hypothesis through empirical observation and experimentation. Moreover, speculative thinkers and futurists are envisioning potential scenarios for the creation of simulated worlds and contemplating the ethical and existential ramifications of such endeavors.

7. Conclusion: Embracing the Mystery of Existence

In the grand tapestry of human knowledge and inquiry, the simulation hypothesis stands as a testament to the boundless curiosity and imagination of the human spirit. Whether our universe is indeed a digital construct or the product of natural processes, the quest to understand the nature of reality transcends the confines of empirical observation and leads us to explore the deepest recesses of our minds and souls. As we navigate the complexities of existence and grapple with the enigma of our own existence, let us embrace the mystery that surrounds us and continue to ponder the profound questions that shape our understanding of the cosmos.

CHAPTER NINETEEN

EXISTENCE OF GHOSTS

Introduction

Ghosts and spirits have long been a subject of fascination and fear in human culture, often depicted in literature, folklore, and media as ethereal beings lingering between the realms of the living and the dead. While traditionally viewed as purely supernatural phenomena, there is a growing body of scientific research and inquiry seeking to understand the potential existence of ghosts and spirits from a rational, empirical perspective. This chapter delves into the scientific evidence surrounding the existence of ghosts and spirits, examining various theories, experiments, and observations that challenge conventional beliefs and offer intriguing insights into the nature of consciousness and the afterlife.

Historical Perspectives and Cultural Beliefs

To understand the contemporary scientific inquiry into ghosts and spirits, it's essential to explore the historical and cultural contexts that have shaped our perceptions of these phenomena. Throughout history, diverse cultures have held beliefs in the existence of spirits, ancestors, and supernatural entities, often attributing unexplained phenomena to the influence of these otherworldly beings. From ancient civilizations to modern religions, tales of ghostly encounters and spectral apparitions have been woven into the fabric of human experience, serving as a source of both comfort and terror.

Psychological Explanations and Paranormal Experiences

One approach to understanding ghosts and spirits from a scientific perspective is through the lens of psychology. Researchers have proposed various psychological theories to explain paranormal experiences, including hallucinations, sleep paralysis, and suggestibility. For example, the phenomenon of sleep paralysis, characterized by a temporary inability to move or speak upon waking or falling asleep, has been linked to vivid hallucinations of ghostly figures and malevolent entities. Studies have shown that factors such as stress, sleep deprivation, and cultural beliefs can influence the likelihood of experiencing paranormal phenomena, highlighting the role of psychology in shaping our perceptions of the supernatural.

Quantum Mechanics and Consciousness

In recent years, some scientists and philosophers have turned to quantum mechanics to explore the nature of consciousness and its potential connection to paranormal phenomena. Quantum theory suggests that at the subatomic level, particles can exist in multiple states simultaneously and can influence each other over vast distances through entanglement. Proponents of this view argue that consciousness may arise from quantum processes within the brain, leading to the possibility of consciousness surviving death in some form. While speculative, this hypothesis offers a new framework for understanding the relationship between the mind, the body, and the universe, opening up intriguing avenues for further research into the existence of ghosts and spirits.

Experimental Investigations and Anomalous Phenomena

Despite the skepticism of mainstream science, there have been numerous reports of anomalous phenomena that defy conventional explanation, ranging from ghostly apparitions and poltergeist activity to electronic voice phenomena (EVP) and near-death experiences (NDEs). While individual anecdotes are not sufficient to establish scientific proof, some researchers have conducted controlled experiments and investigations to study these phenomena under controlled conditions. For example, parapsychologists have conducted experiments using instruments such as electromagnetic field (EMF) detectors and thermal imaging cameras to detect anomalies associated with haunted locations. While the results of such studies are often inconclusive, they provide valuable data for further exploration and hypothesis testing.

Albert Einstein's beliefs regarding the existence of ghosts add an intriguing dimension to the scientific discourse surrounding this topic. Einstein, renowned for his groundbreaking work in physics, including the theory of relativity and the concept of mass-energy equivalence ($E=mc^2$), once famously stated, "Energy cannot be created nor destroyed; it can only be changed from one form to another."

Einstein's adherence to the conservation of energy principle reflects his profound understanding of the fundamental laws governing the universe. While Einstein's beliefs about ghosts are not extensively documented, it is known that he held a deep interest in philosophical and metaphysical questions beyond the scope of conventional science.

One interpretation of Einstein's statement is that the energy associated with consciousness, emotions, and human experience may persist beyond the physical body, potentially contributing to the existence of ghosts or spirits. From this perspective, ghosts could be regarded as manifestations of energy patterns or information imprinted on the fabric of spacetime, echoing Einstein's conception of the interconnectedness of matter and energy.

While Einstein's views on ghosts may have been speculative and philosophical rather than empirical, his acknowledgment of the mysterious nature of energy and its potential implications for the afterlife adds an intriguing layer to the ongoing scientific exploration of paranormal phenomena. In the quest to unravel the mysteries of the universe, Einstein's legacy serves as a reminder of the boundless possibilities that lie beyond the boundaries of our current understanding.

Conclusion

In conclusion, the existence of ghosts and spirits remains a subject of scientific debate and inquiry, with researchers from diverse disciplines seeking to unravel the mysteries of the paranormal. While skepticism and caution are warranted in evaluating extraordinary claims, the quest to understand the nature of consciousness and its potential transcendence beyond the physical realm continues to inspire curiosity and exploration. Whether ghosts and spirits exist as objective entities or as manifestations of the human psyche, the search for truth and meaning in the face of the unknown remains an enduring pursuit of the human spirit.

CHAPTER TWENTY

DARK MATTER AND ENERGY

- **Dark matter**

Dark matter is a fascinating enigma that pervades the cosmos, constituting approximately 27% of the universe's total mass-energy content. Despite its ubiquitous presence, dark matter remains hidden from our direct observation, as it neither emits nor absorbs light, earning it the apt moniker "dark."

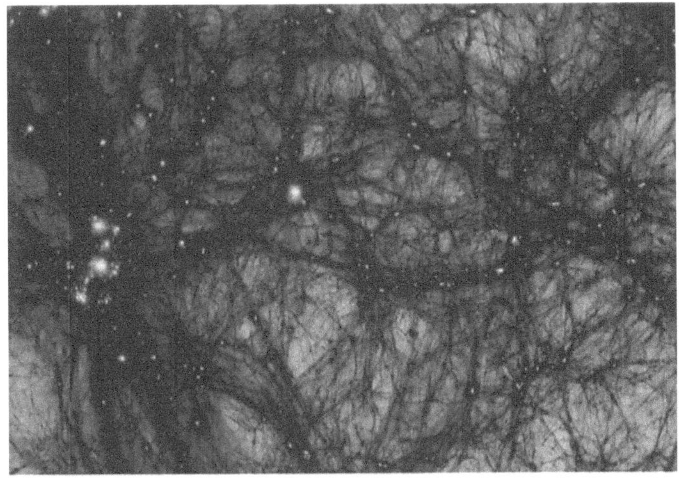

Our awareness of dark matter stems primarily from its gravitational influence on visible matter, radiation, and the cosmic web of galaxies. It exerts a gravitational pull that affects the motions of stars and galaxies on cosmic scales, shaping the very fabric of the universe.

While dark matter's gravitational footprint is undeniable, its fundamental nature eludes scientific scrutiny. It interacts weakly, if at all, with ordinary matter, rendering it extraordinarily challenging to detect. This elusiveness has sparked a quest to unravel its mysteries, driving a multifaceted approach encompassing astrophysical observations, theoretical modeling, and experimental endeavors.

A prevailing hypothesis posits that dark matter consists of exotic particles yet to be discovered by particle physics experiments. These hypothetical particles, such as Weakly Interacting Massive Particles (WIMPs) or Axions, would only interact with regular matter via gravity and possibly the weak nuclear force, evading detection by conventional means.

Numerous experiments, both terrestrial and space-based, are underway worldwide, seeking to directly detect dark matter particles or indirectly infer their existence through their subtle interactions with the observable universe. By unraveling the secrets of dark matter, scientists aim to deepen our understanding of the cosmos and unlock the mysteries of its hidden realms.

- DARK ENERGY

Dark energy is a mysterious force that permeates the universe and drives its accelerated expansion. It represents approximately 68% of the universe's total energy density. Unlike dark matter, which exerts gravitational attraction and slows down the expansion of the universe, dark energy behaves in the opposite manner, causing the expansion to accelerate.

The discovery of dark energy's existence came from observations of distant supernovae in the late 1990s. These observations revealed that the rate of expansion of the universe is increasing over time, contrary to what was expected based on the gravitational interactions of visible matter and dark matter.

The exact nature of dark energy remains one of the most significant unanswered questions in modern cosmology. One prevalent hypothesis is that dark energy corresponds to the energy associated with empty space, often referred to as the cosmological constant. According to Einstein's theory of general relativity, empty space can possess energy that exerts a repulsive gravitational force, causing the accelerated expansion of the universe.

Alternatively, dark energy could arise from other exotic phenomena or modifications to Einstein's theory of gravity on cosmological scales. Understanding dark energy is crucial for elucidating the ultimate fate of the universe and refining our comprehension of fundamental physics on the largest scales. Ongoing astronomical surveys and experiments aim to probe the nature of dark energy and unravel its profound implications for the cosmos.

CHAPTER TWENTY-ONE

PLANETS LIKE EARTH

Scientists have identified numerous such planets, some with simple life forms and complex ecosystems. Factors like atmospheric pressure, temperature, magnetic fields, and planet size influence their potential to support life. Two specific planets, 5715.1 and 5554.1, have stable atmospheres, suitable temperatures, and water, making them potential candidates for life. The first, 5715.1, has a lower surface temperature and longer orbit, keeping it in the habitable zone for an extended period. The second, 5554.1, has a surface temperature suitable for human life and a large amount of water. Scientists are currently investigating these planets further, and more discoveries are expected. Methods like transit and radial velocity have allowed the identification of numerous potentially habitable planets, expanding the search for extraterrestrial life. One such planet, 55 Cancri e, has a significant amount of water and essential materials for life, and messages were sent to it by NASA and Met Office in 2017 and 2018. While there are challenges, such as Jupiter's radiation, the potential for discovering more super-habitable planets and atmospheric bottles with a gravitational pull is vast.

They mention that these planets, such as the one in the Antares system, have environments that could make humans feel superpowers and even extend their lifespan up to 50 billion years. The researchers have discovered numerous such planets, some of which have simple life forms and even complex ecosystems. These planets are located in various star systems, and their sizes and orbits influence their potential to support life. The team has also considered factors like atmospheric pressure, temperature, and magnetic fields to determine the habitability of these planets. They emphasize that while there is no universal language barrier, learning a new language is essential for professional and job opportunities. They encourage downloading the Duolingo app to learn new languages and explore various cultures. The researchers also mention that the size and mass of a planet, along with its orbit, can significantly impact its potential to support life. They conclude by stating that while the concept is basic, the potential implications are vast, as larger planets with heavier tablesurfaces and more mass could lead to more significant discoveries.

scientists discuss their discovery of two potentially habitable planets, 5715.1 and 5554.1. These planets have several favorable factors for potential life, including a stable atmosphere, a suitable temperature range, and the presence of water. The first planet, 5715.1, has a surface temperature lower than Earth's, and its orbit around its star is longer than Earth's, keeping it in the habitable zone for a more extended period. The second planet, 5554.1, has a surface temperature suitable for human life and is located in the habitable zone of its star. Additionally, this planet has a large amount of water and is believed to have a significant amount of nitrogen, which is essential for life. Researchers are currently investigating these planets further, and more discoveries are expected in this category. The first planet, Elas 1140, was discovered in 2017 using a transit method, and it is located in the habitable zone of a red dwarf star. This method involves observing the transit of a planet across its star's face, and if the planet's size and orbit are such that it blocks a significant portion of the star's light, it can be identified as a potential habitable planet. The second planet, Louisine, was discovered using radial velocity methods, which involve measuring the wobble of a star caused by the gravitational influence of a planet orbiting around it. These methods have allowed scientists to identify numerous potentially habitable planets, expanding the search for extraterrestrial life.

Observer discovered that this planet has a significant amount of water and essential materials for life such as ammonia and hydrocarbons. This short planet, named 55 Cancri e, is considered perfect and may even have an atmosphere that could support human life. NASA and Met Office sent messages to this planet in 2017 and 2018, and one message contained short musical composition and the other contained a scientific tutorial about the university's research. If Louis Pasteur was an inhabitant of this planet, they could easily breathe and live there. However, there is a small problem with Jupiter's radiation. If we ignore this issue, other super-habitable planets and a great atmospheric bottle with a gravitational pull could exist.

Kepler-22b

In the vast expanse of the cosmos, distant stars harbor a myriad of worlds, each with its own unique characteristics and potential for life. Among these distant realms lies Kepler-22b, an exoplanet located within the habitable zone of its parent star, Kepler-22. Discovered by NASA's Kepler Space Telescope in 2011, Kepler-22b has captured the imagination of scientists and astronomers around the world as a prime candidate for hosting liquid water and potentially harboring life.

Discovery and Characteristics

Kepler-22b was detected using the transit method, where astronomers observe the dimming of a star as a planet passes in front of it, causing a temporary decrease in brightness. Located approximately 600 light-years away in the constellation of Cygnus, Kepler-22b orbits its parent star at a distance roughly equivalent to that of Earth from the Sun. With a radius estimated to be about 2.4 times that of Earth, Kepler-22b is classified as a "super-Earth," indicating that it is larger than our home planet but smaller than gas giants like Jupiter.

Habitability and Climate

One of the most intriguing aspects of Kepler-22b is its location within the habitable zone of its parent star, where conditions may be suitable for the existence of liquid water on its surface. The habitable zone, also known as the "Goldilocks zone," is the region around a star where temperatures are neither too hot nor too cold for water to exist in liquid form, a crucial ingredient for life as we know it. While the precise composition and climate of Kepler-22b remain uncertain, computer models suggest that it could have a dense atmosphere and a moderate climate, making it a potentially habitable world.

Challenges and Future Exploration

Despite its tantalizing potential, studying Kepler-22b presents numerous challenges for astronomers. Its considerable distance from Earth makes detailed observations and direct imaging challenging with current technology. Furthermore, the composition of its atmosphere and surface characteristics are still largely unknown, leaving many questions unanswered about the planet's potential habitability.

Nevertheless, astronomers are optimistic about the future of exoplanet research and the prospects for studying worlds like Kepler-22b in greater detail. Advanced telescopes and space missions, such as the James Webb Space Telescope and the upcoming Nancy Grace Roman Space Telescope, promise to revolutionize our understanding of distant exoplanets and their potential for hosting life. By studying the atmospheres of exoplanets through techniques like transit spectroscopy, scientists hope to detect biosignatures—chemical signatures indicative of life—that could provide tantalizing clues about the existence of extraterrestrial life beyond our solar system.

Kepler-22b stands as a beacon of hope in the search for life beyond our solar system, a distant world that beckons us to explore its mysteries and unlock the secrets of our cosmic neighbors. While much remains to be discovered about this enigmatic exoplanet, its location within the habitable zone and its Earth-like characteristics make it a compelling target for future exploration. As humanity ventures forth into the cosmos, Kepler-22b serves as a reminder of the boundless possibilities that await us among the stars.

CHAPTER TWENTY-TWO

MULTIVERSE

The concept of a multiverse stretches the bounds of our imagination, proposing the existence of a vast ensemble of universes beyond our own, each potentially with its own distinct laws of physics, constants, and cosmic properties. This idea arises from the confluence of various theoretical frameworks in cosmology, quantum mechanics, and string theory, pushing the boundaries of our understanding of reality to the limit.

One avenue through which the multiverse hypothesis emerges is the framework of inflationary cosmology. According to this theory, the rapid expansion of space that occurred in the early moments of the universe's history may have given rise to a plethora of "pocket" or "bubble" universes, each birthing its own unique cosmic tapestry. These universes, though existing side by side in an unimaginably vast cosmic landscape, would remain causally disconnected from one another, evolving according to their own internal dynamics.

In the realm of string theory, which seeks to unify all fundamental forces and particles in the universe, the notion of a multiverse finds fertile ground. String theory postulates the existence of additional spatial dimensions beyond the familiar three dimensions of space and one dimension of time. Within this framework, the different ways these extra dimensions can be compactified could give rise to a dizzying array of universes, each characterized by its own particular configuration of dimensions and physical laws.

Despite its theoretical appeal, the multiverse hypothesis remains firmly ensconced in the realm of speculation, as direct empirical evidence for the existence of other universes eludes our current observational capabilities. Yet, the prospect of a multiverse tantalizes scientists and philosophers alike, challenging us to rethink our place in the cosmos and confront the profound mysteries that lie beyond the horizon of our cosmic comprehension. As we continue to explore the frontiers of theoretical physics and observational cosmology, the question of whether a multiverse truly exists stands as one of the most profound and tantalizing enigmas of our age.

Delving deeper into the multiverse hypothesis, we encounter a tapestry of theoretical constructs that weave together the fabric of reality in ways that stretch the limits of our comprehension. One such intriguing manifestation of the multiverse is the concept of the "many-worlds" interpretation in quantum mechanics.

According to the many-worlds interpretation, every quantum event results in the branching of reality into multiple parallel universes, each representing a different outcome of that event. This interpretation suggests that every conceivable possibility allowed by the laws of quantum mechanics plays out in its own separate universe, encompassing an infinite array of alternate realities.

Furthermore, cosmic inflation, the rapid expansion of the universe in its infancy, provides fertile ground for the proliferation of diverse universes within a grand cosmic tapestry. Inflationary models posit that regions of space can undergo exponential expansion, giving rise to "pocket" universes with distinct properties, effectively creating a multiverse on an unimaginably vast scale.

Moreover, recent developments in cosmology and theoretical physics, such as the holographic principle and quantum entanglement, hint at deeper connections between seemingly disparate universes within the multiverse. The holographic principle suggests that all the information in a three-dimensional space can be encoded on its two-dimensional boundary, implying a profound interconnectedness between different regions of space-time.

While the multiverse remains speculative and beyond current observational reach, it serves as a fertile ground for theoretical exploration and philosophical contemplation. The idea challenges our notions of reality, inviting us to ponder the nature of existence and our place within the vast cosmic expanse.

As we continue to push the boundaries of human understanding through theoretical inquiry and observational exploration, the multiverse stands as a tantalizing frontier, beckoning us to unravel its mysteries and unlock the secrets of the cosmos. Whether the multiverse exists as a concrete reality or as a mathematical abstraction, its exploration promises to deepen our appreciation of the profound beauty and complexity of the universe in which we dwell.

References

- Sagan, C. (1994). Pale Blue Dot: A Vision of the Human Future in Space. New York: Random House.
- Hawking, S. (1988). A Brief History of Time: From the Big Bang to Black Holes. New York: Bantam Books.
- Tyson, N. D. (2017). Astrophysics for People in a Hurry. New York: W.W. Norton & Company.
- Greene, B. (2011). The Hidden Reality: Parallel Universes and the Deep Laws of the Cosmos. New York: Vintage Books.
- Smolin, L. (2001). Three Roads to Quantum Gravity. New York: Basic Books.
- Further Reading
- Cosmos by Carl Sagan
- The Elegant Universe by Brian Greene
- Black Holes and Time Warps: Einstein's Outrageous Legacy by Kip S. Thorne
- The Universe in a Nutshell by Stephen Hawking
- Welcome to the Universe: An Astrophysical Tour by Neil deGrasse Tyson, Michael A. Strauss, and J. Richard Gott
- Einstein's Theory of Special Relativity 1905.
- H.G WELLS (TIME MACHINE)